Reinforced Concrete Structures: Analysis, Drawing and Design

Reinforced Concrete Structures: Analysis, Drawing and Design

Ezra Bawden

www.willfordpress.com

Published by Willford Press,
118-35 Queens Blvd., Suite 400,
Forest Hills, NY 11375, USA

ISBN: 978-1-64728-503-6

Cataloging-in-Publication Data

Reinforced concrete structures : analysis, drawing and design / Ezra Bawden.
p. cm.
Includes bibliographical references and index.
ISBN 978-1-64728-503-6
1. Reinforced concrete construction. 2. Structural design.
3. Reinforced concrete construction--Standards.
I. Bawden, Ezra.

TA683.2 .R45 2023
624.183 41--dc23

For information on all Willford Press publications
visit our website at www.willfordpress.com

Contents

Preface

Reinforced concrete (RC) refers to a type of building material that combines two or more materials with different physical properties to impart greater tensile strength and ductility to the structure of a building. RC structures are made up of composite materials constituted by concrete material, composites or polymers, and steel bars. The various methods utilized in the design of RC structures include the limit state method (LSM), the working stress method (WSM), and the ultimate load method (ULM). There are various types of structures that can be constructed using RC such as floating structures, marine structures, flyovers, chimneys and towers, water tanks, and retaining walls. This book is compiled in such a manner, that it will provide in-depth knowledge about the drawing, design, and analysis of reinforced concrete structures. It is appropriate for students seeking detailed information in this area of civil engineering as well as for experts.

After months of intensive research and writing, this book is the end result of all who devoted their time and efforts in the initiation and progress of this book. It will surely be a source of reference in enhancing the required knowledge of the new developments in the area. During the course of developing this book, certain measures such as accuracy, authenticity and research focused analytical studies were given preference in order to produce a comprehensive book in the area of study.

This book would not have been possible without the efforts of the authors and the publisher. I extend my sincere thanks to them. Secondly, I express my gratitude to my family and well-wishers. And most importantly, I thank my students for constantly expressing their willingness and curiosity in enhancing their knowledge in the field, which encourages me to take up further research projects for the advancement of the area.

Ezra Bawden

1

Layout Drawing

1.1 General Layout of Buildings

Drawing is the language of engineers. An engineer must be well conversant with drawings. Drawings represent reduced shape of structure and the owner will be able to see what is going to happen.

Drawings are prepared as per the requirements of owner. In case of public buildings, the functional aspects are studied and accordingly the drawings are prepared as per recommendations laid down in National Building Code (N.B.C) or as per Indian Standard specifications.

Any modifications like additions or omissions can be suggested from a study of the drawings before actual construction of the structure is started. Drawings provide a language with specific data to Architects, Engineers and workmen at the site to construct the structure accordingly.

In case of public buildings or any other civil engineering works, it is essential to work out different items of construction with their quantities for estimating the total cost of construction project. For this purpose, drawings of different parts and different views are essential so that the approval of work from the sanctioning authority can be obtained.

Further, the detailed drawings form an essential contract documents, when the work is handed over to a contractor. Hence it is necessary to prepare detailed drawings, which will inform the contractor, the exact information, which he needs during the construction of different items of work.

Drawings, thus prepared should be carefully even after the completion of work. Thus, it becomes asses the possibility of further vertical expansion by referring to the foundation details initially provided.

Requirements of good drawing:

- Drawing should be clear, simple and clean.
- Should agree with the actual measurements by the accurately drawn scaled measurements.

- Exact information should be provided in order to carry out the work at site without scaling for missing measurements.
- Only minimum notes to support the drawings should be indicated in the drawings.
- Sufficient space should be provided between the views so as to mark the dimensions without crowding.

Conventional Signs and Symbols

Conventional signs are used to represent the particular item like stone masonry, brick masonry, concrete etc. in the section of drawing. (i.e.,) when the materials are cut by any imaginary plane.

Conventional symbols are provided to indicate doors, windows, their fixing, and movement of shutters. When they are closed or opened, various water supply and sanitary fixtures like tap, wash basin, urinals, Kitchen sink, shower etc., and symbols are used to indicate the position of electrical fittings like lamp, switch, power socket, fan etc. To indicate positions of furniture on drawing room, bedroom, suitable symbols are used.

The conventional signs for civil engineering materials are shown in below figure:

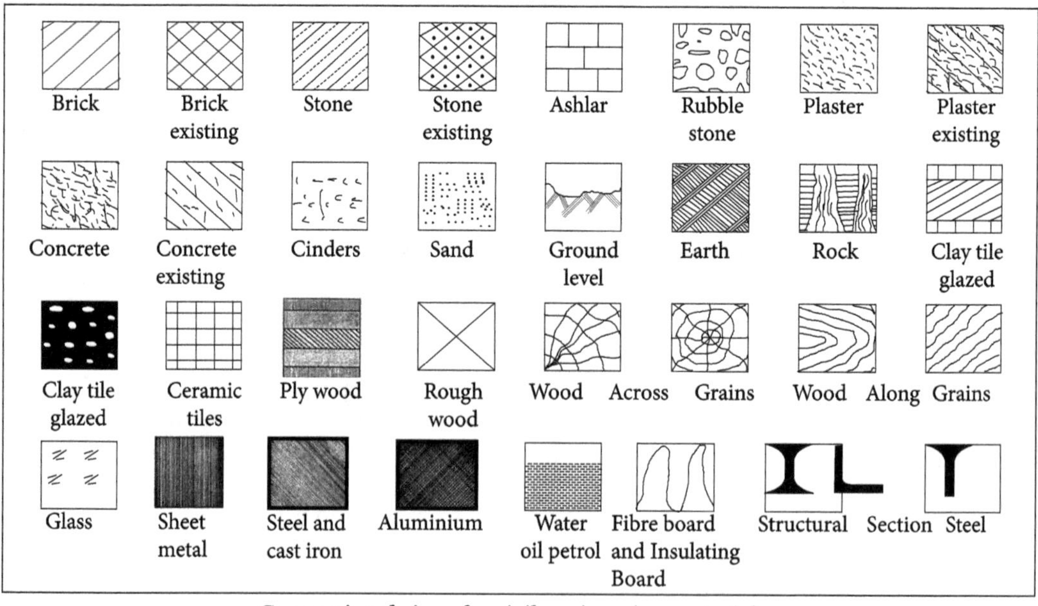

Conventional signs for civil engineering materials.

The Bureau of Indian standards (B.I.S) has recommended the conventional signs and symbols for the following purposes:

- Avoid confusion and to understand the drawings.
- Save the time in making out various details in the drawing.

- Identify the various details of materials, Electrical fixtures, water supply and sanitary fittings, Position of furniture's etc.
- To prevent any dispute between contractor and owner in the actual construction of the structure.

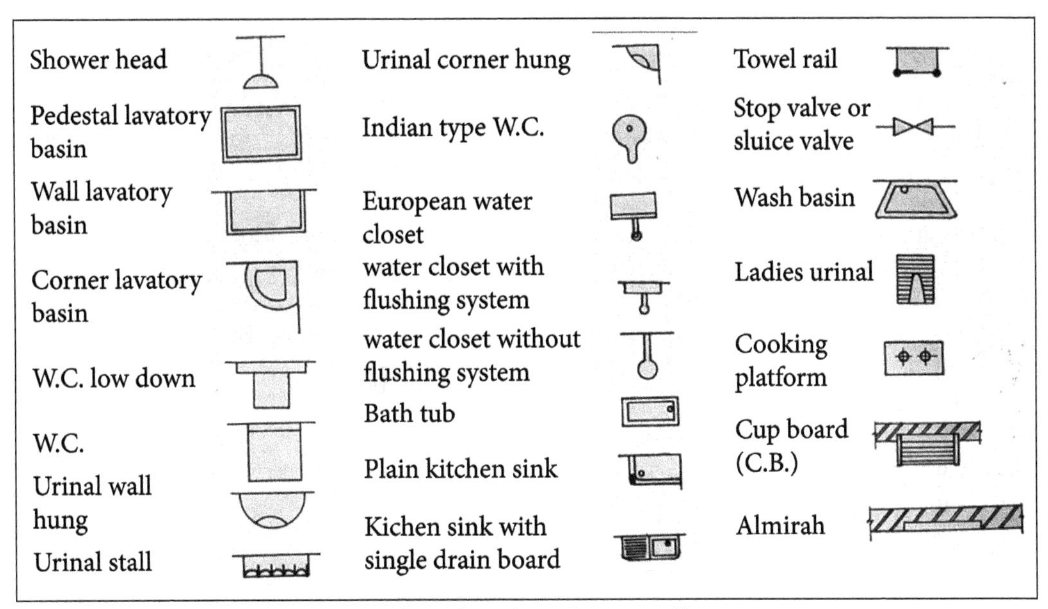

Water supply and Sanitary fixtures – diagrams.

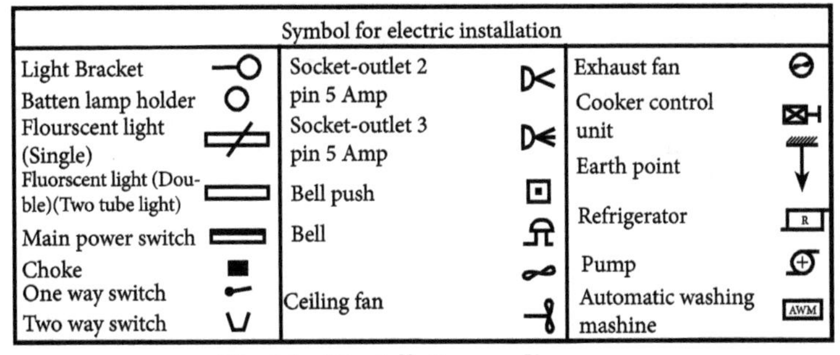

Electrical Installations – diagrams.

General Layout of Building

- Positioning and orientation of columns: Following are some of the building principles, which help in deciding the columns positions.
- Columns should preferably be located at or near the corners of a building and at the intersection of beams/walls.
- Select the position of columns so as to reduce bending moments in beams.
- Avoid larger spans of beams.
- Avoid larger centre-to-centre distance between columns.

- Columns on property line.

Orientation of Columns

Avoid Projection of Columns

The projection of columns outside the wall in the room should be avoided as they not only give bad appearance but also obstruct the use of floor space, creating problems in placing furniture flush with the wall. The width of the column is required to be kept not less than 200mm to prevent the column from being slender.

The spacing of the column should be considerably reduced so that the load on column on each floor is less and the necessity of large sections for columns does not arise.

Orient the column so that the depth of the column is contained in the major plane of bending or is perpendicular to the major axis of bending.

This is provided to increase moment of inertia and hence greater moment resisting capacity. It will also reduce Leff/d ratio resulting in increase in the load carrying capacity of the column.

Positioning of Beams

Beams shall normally be provided under the walls or below a heavy concentrated load to avoid these loads directly coming on slabs.

Avoid larger spacing of beams from deflection and cracking criteria. (The deflection varies directly with the cube of the span and inversely with the cube of the depth i.e. L3/D3. Consequently, increase in span L which results in greater deflection for larger span).

Spanning of Slabs

This is decided by supporting arrangements. When the supports are only on opposite edges or only in one direction, then the slab acts as a one way supported slab. When the rectangular slab is supported along its four edges it acts as a one way slab when

$$L_y / L_x < 2.$$

The two way action of slab not only depends on the aspect ratio but also on the ratio of reinforcement on the directions. In one way slab, main steel is provided along with short span only and the load is transferred to two opposite supports. The steel along the long span just acts as the distribution steel and is not designed for transferring the load but to distribute the load and to resist shrinkage and temperature stresses.

A slab is made to act as a one way slab spanning across the short span by providing main steel along the short span and only distribution steel along the long span. The provision of more steel in one direction increases the stiffness of the slab in that direction.

According to elastic theory, the distribution of load being proportional to stiffness in two orthogonal directions, major load is transferred along the stiffer short span and the slab behaves as one way. Since, the slab is also supported over the short edge there is a tendency of the load on the slab by the side of support to get transferred to the nearer support causing tension at top across this short supporting edge.

Since, there does not exist any steel at top across this short edge in a one way slab interconnecting the slab and the side beam, cracks develop at the top along that edge. The cracks may run through the depth of the slab due to differential deflection between the slab and the supporting short edge beam/wall. Therefore, care should be taken to provide minimum steel at top across the short edge support to avoid this cracking.

A two way slab is generally economical compare to one way slab because steel along both the spans acts as main steel and transfers the load to all its four supports. The two way action is advantageous essentially for large spans (>3m) and for live loads $(> 3kN/m^2)$. For short spans and light loads, steel required for two way slabs does not differ appreciably as compared to steel for two way slab because of the requirements of minimum steel.

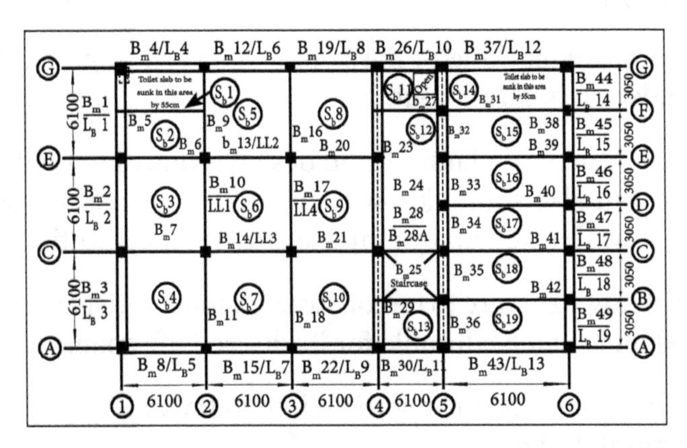

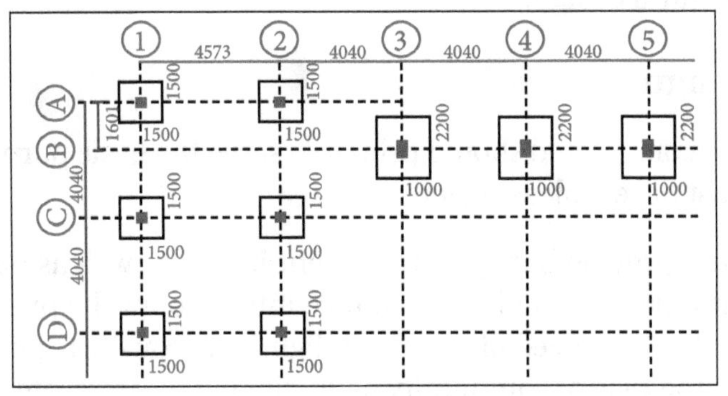

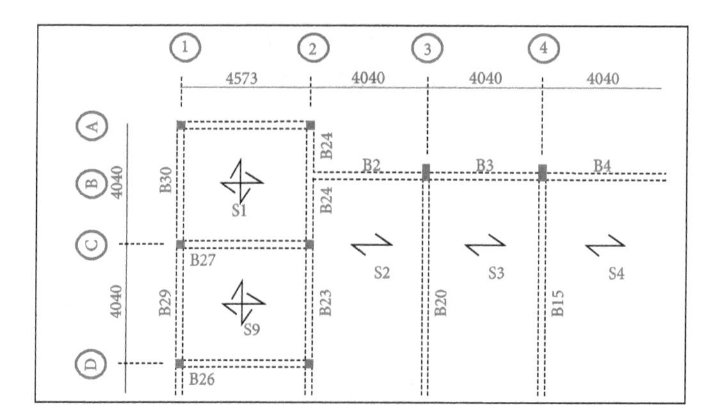

Procedure for Layout of Building

- The layout of the building is the first stage before the execution of the project.
- The layout is carried on every floor which is important for the casting of various units of the floor.
- Layout is the fixing of center line for column and beams and other structure i.e. lift etc.
- Two reference points is used to make horizontal grid and one reference point is used for vertical grid.
- After the complete marking of horizontal and vertical grids, the reinforcement for the foundation is carried out.

The entire process of structural planning and design requires not only imagination and conceptual thinking but also sound knowledge of practical aspects such as recent design codes and bye-laws, backed up by ample experience, institution and judgment.

It is emphasized that any structure to be constructed must satisfy the need efficiency for which it is intended and shall be durable for its desired life span.

Thus the design of any structure is categorizes into following two main types: Functional design and Structural design.

Functional Design

The structure to be constructed should primarily serve the basic purpose for which it is to be used and must have a pleasing look.

The building should provide happy environment inside as well as outside. Therefore, the functional planning of a building must take into account the proper arrangements of room/halls to satisfy the need of the client, lighting, acoustics, good ventilation, unobstructed view in the case of community halls, cinema theaters, etc.

Structural Design

Once the form of the structure is selected, the structural design process starts. Structural design is an art and science of understanding the behavior of structural members subjected to loads and designing them with economy and elegance to give a safe, serviceable and durable structure.

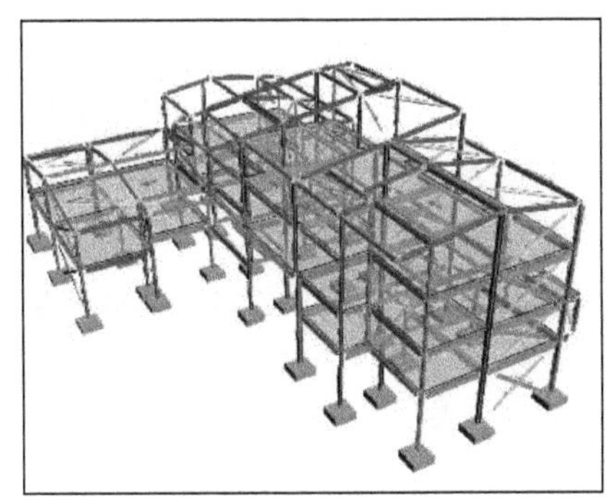

Structural Design.

Footing

The type of footing depends upon the load carried by the column and the bearing capacity of the supporting soil. The soil under the foundation is more susceptible to large variations. Even under one small building the soil may vary from soft clay to a hard murum.

The nature and properties of soil may change with season and weather, like swelling in wet weather. Increase in moisture content results in substantial loss of bearing capacity in case of certain soils which may lead to differential settlements.

It is necessary to conduct the survey in the areas for soil properties. For framed structure, isolated column footings are normally preferred except in case of exists for great depths, pile foundations can be an appropriate choice.

If columns are very closely spaced and bearing capacity of the soil is low, raft foundation can be an alternative solution. For a column on the boundary line, a combined footing or a raft footing may be provided.

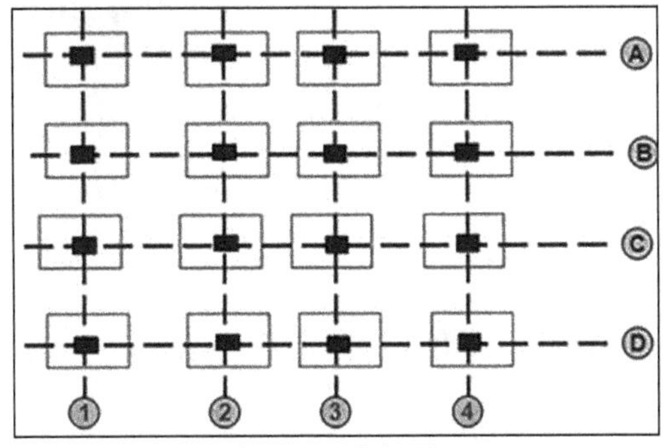

Column and footing.

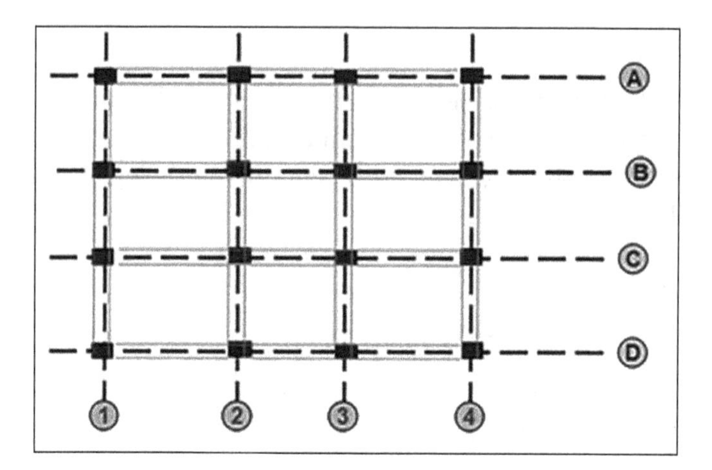

Assumptions

The following are the assumptions made in the earthquake resistant design of structures:

- Earthquake causes impulsive ground motions, which are complex and irregular in character, changing in period and amplitude each lasting for small duration. Therefore resonance of the type as visualized under steady-state sinusoidal excitations, will not occur as it would need time to build up such amplitudes.
- Earthquake is not likely to occur simultaneously with wind or max. Flood or max, sea waves.
- The value of elastic modulus of materials, wherever required, maybe taken as per static analysis.

Problem

1. A clear dimension of factory floor is 11.75 m x 19.75 m, Spacing of columns 4m c/c, Size of columns 250 mm × 450 mm, Span of steel truss is 12.25 m c/c, At the ends to support the gable wall additional two RCC columns of size 250 mm × 450 mm are to be provided at 4m c/c measured from end columns, All the walls all-round are 250 mm thick, Height of columns 3m, Size of footing 1.4m × 1.8m, Thickness of footing 300 mm uniform, Depth of foundation 1.2m below ground level. Let us draw the layout drawing for above data.

Solution:

Given:

- A clear dimension of factory floor is 11.75 m × 19.75 m.
- Spacing of columns 4m c/c.

- Size of columns 250 mm × 450 mm.
- Span of steel truss is 12.25 m c/c.
- At the ends to support the gable wall additional two RCC columns of size 250 mm x 450 mm are to be provided at 4m c/c measured from end columns.
- All the walls all-round are 250 mm thick.
- Size of footing 1.4m × 1.8m.
- Thickness of footing 300 mm uniform.
- Height of columns = 3m.
- Depth of foundation 1.2m below ground level.

Centre Line Dimensions of the Building:

Along X-direction = 12.25 – 0.45 = 11.80 m

Along Z-direction = 20.25 -0.25 = 20.00 m

Preliminary Calculations

External Dimensions of the Building:

Along X-direction = 11.75 + (2 × 0.25) = 12.25 m

Along Z-direction = 19.75 + (2 × 0.25) = 20.25 m

Step 1: Draw the centerline of the building having 11,800mm along X-direction and 20,000 mm along Z-directions.

Step 2: Mark these centre lines as grid lines A and D for lines parallel to Z-directions & grid lines 1 and 6 for lines parallel to X-axis.

Step 3: Measure 4000mm from grid lines A and D to get grid line B and C.

Step 4: Measure 4000mm c/c along Z-axis starting from grid line 1 to get grid lines 2, 3, 4 and 5.

Step 5: Draw rectangular filled box of size 250mm × 400mm at the intersection of gird lines along A and D to indicate the position of column along these grid lines as shown in the figure b.

Step 6: Draw rectangular filled box of size 250mm × 400mm at the intersection of grid lines 1, 6 with B and C respectively as shown in the figure b.

Step 7: Draw rectangles each of size 1400 mm × 1800mm symmetrically with respect to centre of column to indicate excavation marking for all columns. Here the shorter side of this rectangle box is parallel to the shorter side of column as shown in the figure b.

Step 8: Sectional elevation and plan of the footing of column is drawn as shown in below figure.

Key Plan

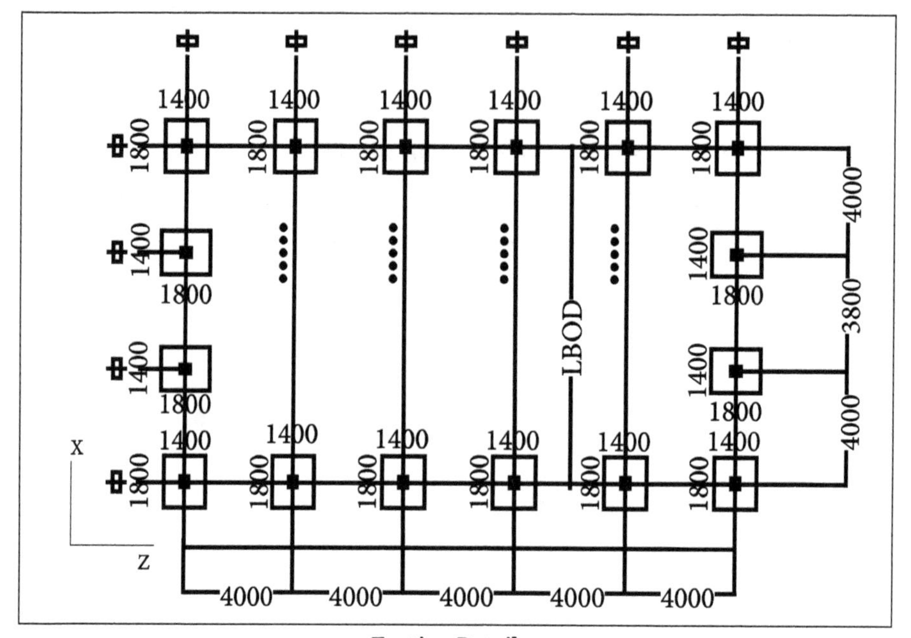

Footing Details.

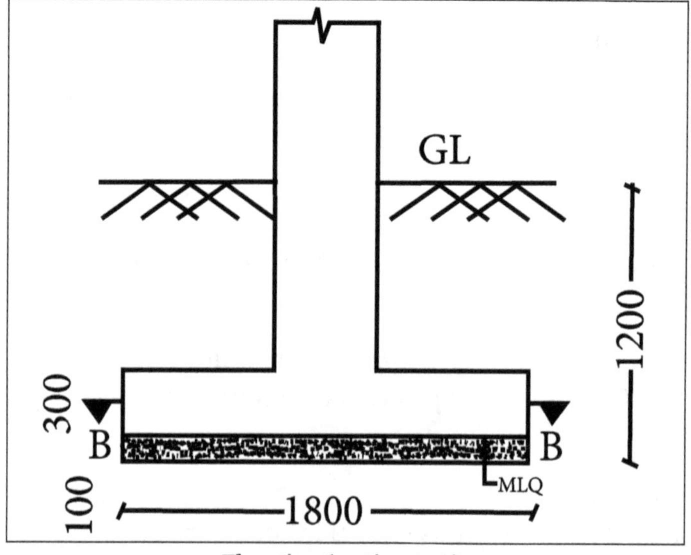

Elevation (section A-A).

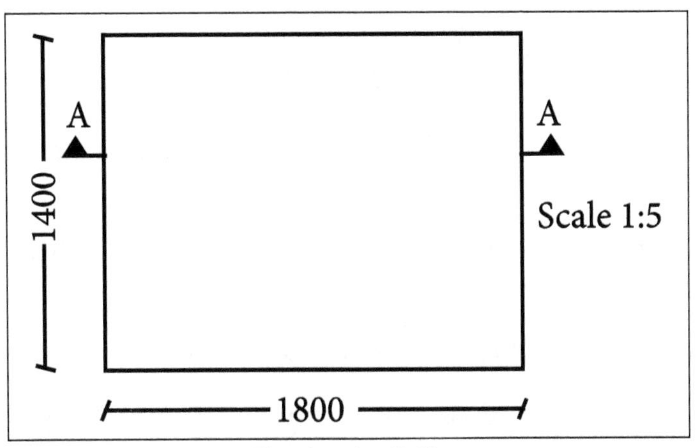

Plan (section B-B).

2. Let us prepare a general layout showing the positions and sizes of Beams and Slabs to a suitable scale for a Community Hall:

- Clear dimension of hall is 12.05 m × 19.8 m.
- Three rows of columns are to be provided along the width with spacing of 6 m c/c.
- Six rows of columns are to be provided along the length with spacing of 4 m c/c.
- All the walls all-round are 200 mm thick.
- Size of columns 200 mm × 450 mm.
- Main beams are of size 200 mm × 600 mm.
- Secondary beams are of size 200 mm × 300 mm.
- Thickness of slab = 150 mm.

Solution:

Given:

Clear dimension of hall is 12.05 m × 19.8 m.

Three rows of columns are to be provided along the width with spacing of 6 m c/c.

Six rows of columns are to be provided along the length with spacing of 4 m c/c.

All the walls all-round are 200 mm thick.

Size of columns 200 mm × 450 mm.

Main beams are of size 200 mm × 600 mm.

Secondary beams are of size 200 mm × 300 mm.

Thickness of slab = 150 mm.

Step 1: Draw the centerline of the building having 12,000 mm along X-direction and 20,000 mm along Z-directions.

Step 2: Mark these centre lines as grid lines A and C for lines parallel to Z-directions & grid lines 1 and 6 for lines parallel to X-axis.

Step 3: Measure 6000mm from grid lines A to get grid line B.

Step 4: Measure 4000mm c/c along Z-axis starting from grid line 1 to get grid lines 2, 3, 4 and 5.

Step 5: Draw rectangular filled box of size 200mm × 450mm at the intersection of gird lines along A and C to indicate the position of column along these grid lines as shown in the figure (a).

Step 6: Draw parallel lines spaced at 200mm symmetrically about the grid lines 1 to 6 to indicate main beams having effective span of 6m as shown in the figure (a).

Step 7: Draw parallel lines spaced 200 mm abutting to end columns to indicate secondary beams as shown in the figure (a). Also draw 200 mm wide secondary beam along grid line B as shown in the figure (a).

Step 8: Mark all the main beams as B1 and Secondary beams as B2.

Step 9: Indicate the slab as one way slab as shown in the figure (a) and mark the slabs as S1.

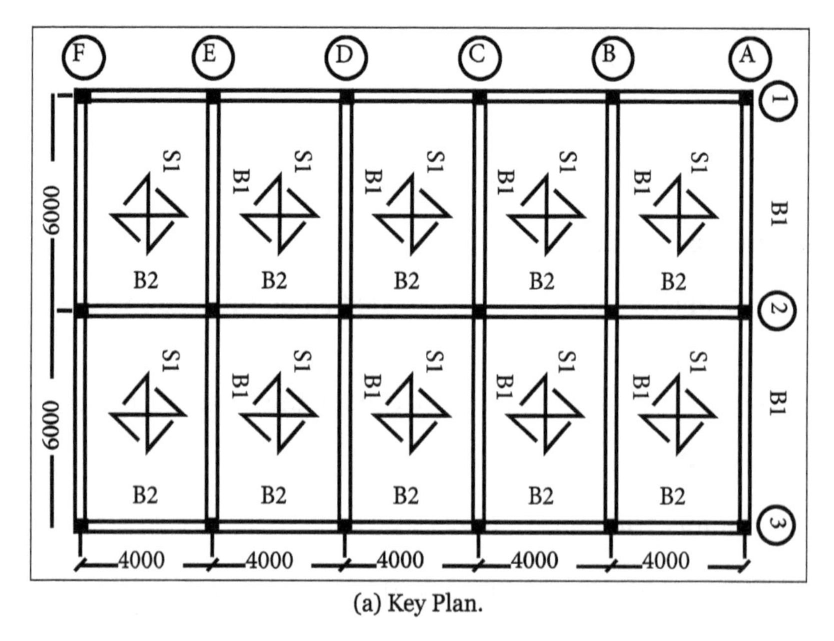

(a) Key Plan.

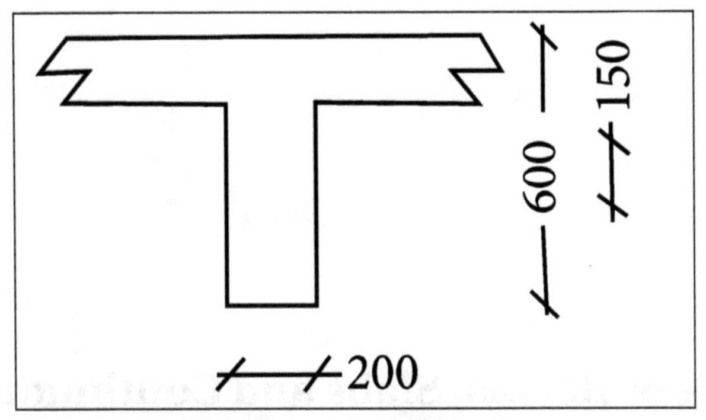

(b) Cross section of beam B1.

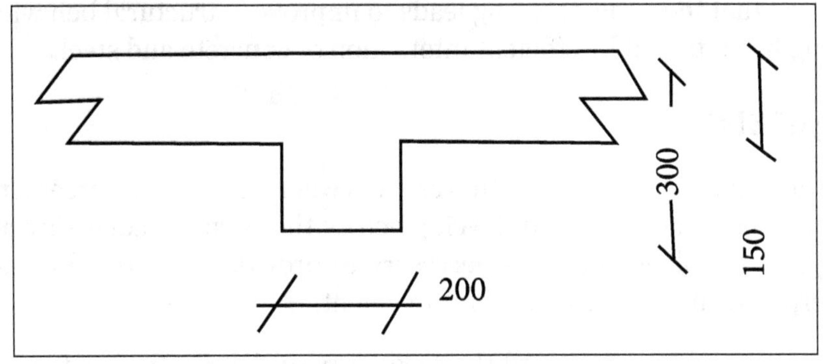

(c) Cross section of beam B2.

2

Detailing of Beam and Slab

2.1 Detailing of Beams, Slabs and Continuous Beams

It is recognized that the good detailing leads to improved structural behaviour and adequate strength is attained by efficient interaction of concrete and steel.

Detailing of Slabs

In the structures like one-way or cantilever slabs where design reinforcement is placed only in one direction cracks could develop across the perpendicular direction due to contraction and shrinkage and it is necessary to provide some reinforcement in the direction perpendicular to the main reinforcement.

This reinforcement is usually called the secondary or distribution or temperature reinforcement. In Case of the members exposed to the sun such as sunshades, canopies, balconies, roof slabs without adequate insulation cover, etc., the minimum reinforcement specified by the Code should be increased by 50 to 100 per cent depending upon the severity of exposure, size of member and local conditions.

Detailing of Beams

Tension reinforcement: The minimum area of tension reinforcement shall not be less than $0.85bd/f_y$, where, b and d are the breadth of the beam or web in case of the T-beam and effective depth respectively. The maximum area of tension reinforcement shall not exceed 0.04bD.

Compression reinforcement: The maximum area of compression reinforcement shall not exceed 0.04bD. Compression steel shall be enclosed by stirrups for the effective lateral restraint.

Anchorage: It is common practice to have flexural bars terminated in the compression zone where the compressive stress acts parallel to the bar. However the compressive stress acting transversely to an anchored bar has been found to be more beneficial. Such a situation is encountered when the reaction acts at the tension face of a beam.

For example, the bottom bars in the end span of a continuous beam show better anchorage at the simply supported end than in the vicinity of the point of contra-flexure where they enter a compression zone. Thus, it is better to seek areas of normal pressure in preference to compression zones for anchorage of flexural reinforcement.

The hooked anchorage at the simply supported end of a beam can be much improved if the hooks are tilted or lie in a near horizontal position. The bars can also be terminated in the tension zone of the beam which contains sufficient transverse web reinforcement.

Concentrated and multilayered arrangement of negative moment (top) reinforcement deteriorates the bond, resulting in an increase of the crack widths. The negative moment steel can be spread into the adjoining monolithic slab preferably by using smaller diameter bars, as shown in the figure (1).

The arrangement will give a slightly larger lever arm and provide better access for vibrators in a usually overcrowded beam-column joint. However, the major part of the flexural reinforcement must be within the multi-legged cage of stirrups.

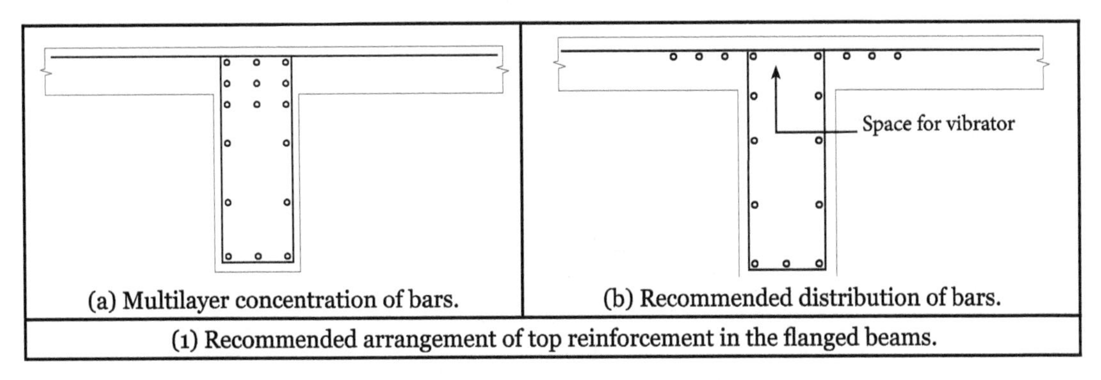

(a) Multilayer concentration of bars.

(b) Recommended distribution of bars.

(1) Recommended arrangement of top reinforcement in the flanged beams.

Shear reinforcement: The longitudinal bars passing through the corners of the stirrups are essential because they must distribute the concentrated bearing received from stirrups in tension. The stirrups should fit tightly and be in contact with the longitudinal bars that they surround.

The normal practice is to bend stirrups around longitudinal bars with an angle of 135°. Some codes permit a 90° turn for stirrups. When shearing forces are present and more than two bars are used to resist flexure, it is desirable to form a truss at each of the longitudinal bars by using multi-legged stirrups.

In the absence of vertical stirrup legs, the centre bars are incapable of resisting vertical forces and thus are inefficient in carrying the bond forces.

Curtailment of flexural steel: The curtailment of flexural bars in the tension zone creates a discontinuity which may result in a sudden increase in tension strain, causing cracks.

It is therefore, essential that the shear strength of such areas of a beam be supplemented by the web reinforcement in the form of additional stirrups in the vicinity of cut-off points of flexural reinforcement in the tension zone.

Contrary to general belief, the bent-up bars from the flexural reinforcement are often responsible for inferior performance.

Side reinforcement: When the depth of the web in a beam exceeds 750 mm, side face reinforcement shall be provided along the two faces. The total area of such reinforcement shall not be less than 0.10% of the web area and shall be distributed equally on the two faces at a spacing not exceeding 300 mm or the web width, whichever is less as shown in the figure.

To guard against the bar yielding locally at crack, in BS: 8110, the spacing of the side reinforcement is limited to 250 mm and the diameter of the bars is restricted to that given by the following expression:

$$\phi = \sqrt{\frac{S_b b}{f_y}}$$

Where,

S_b – Vertical spacing (mm).

b – Width of the beam (mm).

f_y – Characteristic strength (MPa).

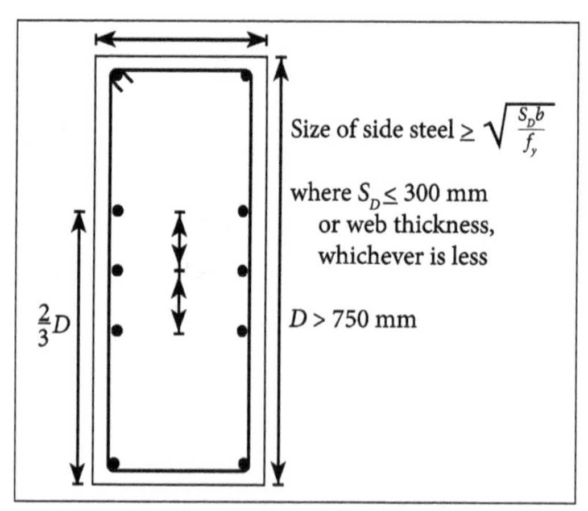

Spacing of side reinforcement.

Detailing of Continuous Beams

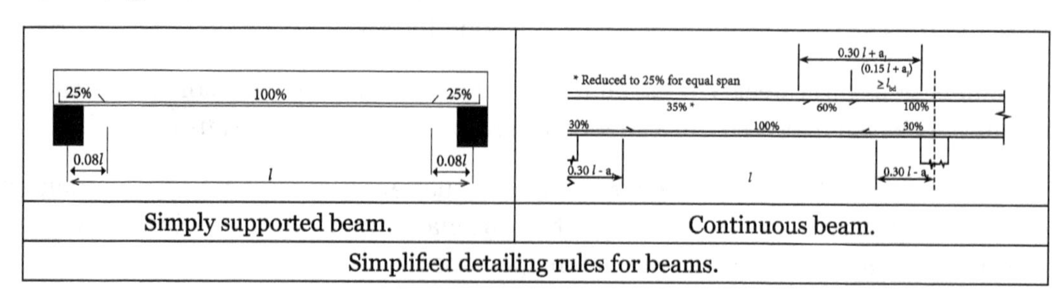

Simply supported beam.	Continuous beam.

Simplified detailing rules for beams.

Where,

l - Effective length.

a_1 - Distance to allow for tensile force due to shear force = z cotθ/2. Can conservatively taken as 1.125d.

l_{bd} - Design anchorage length.

$q_k \leq g_k$.

Minimum of two spans required.

Applies to uniformly distributed loads only.

The shortest span must be greater than or equal to 0.85 times the longest span.

Applies where 15% redistribution has been used.

Limit State Design for Flexure

Singly Reinforced Beams

In singly reinforced the simply supported beams or slabs reinforcing the steel bars are placed near the bottom of the beam or slabs where they are most effective in resisting the tensile stresses.

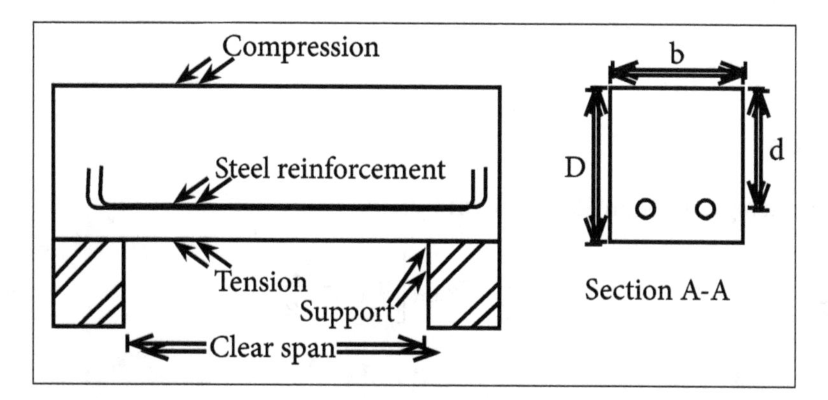

Minimum Effective Depth with Maximum Steel

When b is given or assumed and the effective depth d is unknown, then the section can be designed to have a minimum depth d_{min} by putting $A_s = A_{smax}$. This design shall require a very high steel content.

Until a very shallow depth is essential, the use of A_{smax} is not economical and it is better to use a deep section with less steel. Although, the deflections of a beam with the minimum possible depth may be excessive and need to be checked.

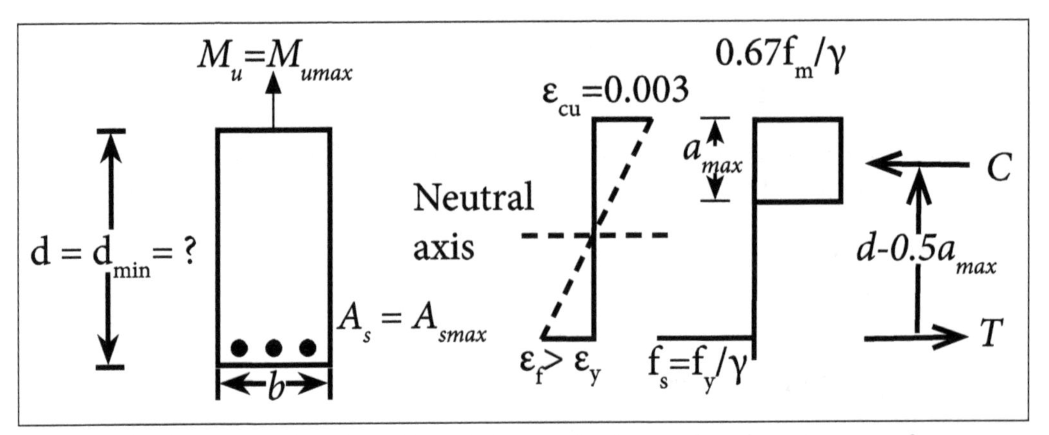

Single reinforced section with a minimum effective depth and maximum steel.

Design Equations

The Figure indicates, the depth will be a minimum, $d = d_{min}$, when A_s is the maximum allowed, A_{smax}.

First calculate d_{min} from,

$$M_u = M_{u\,max} = R_{max} \frac{f_{cu}}{\gamma_c} bd^2_{min}$$

and A_{smax} from,

$$A_{smax} = \mu_{max}\, bd_{min}$$

Design Aids

The ultimate design moment M_u is given by,

$$M_u = C\left[d_{min} - \frac{a_{max}}{2}\right] = 0.67\frac{f_{cu}}{\gamma_c} ba_{max}\left[d_{min} - \frac{a_{max}}{2}\right] \qquad ...(1)$$

Or

$$M_u = T\left[d_{min} - \frac{a_{max}}{2}\right] = A_{s\,max}\frac{f_y}{\gamma_s}\left[d_{min} - \frac{a_{max}}{2}\right] \qquad ...(2)$$

and on substituting,

$$\frac{M_u}{bd^2_{min}} = \frac{1}{K^2_{1\,min}} \text{ and } \frac{f_y}{\gamma_s}\left[1 - \frac{a_{max}}{2d}\right] = K_{2\,min}$$

Equations 1 and 2 become,

$$d_{min} = K_{1min}\sqrt{\frac{M_u}{b}}$$

and

$$A_{s\,max} = \frac{M_u}{K_{2min}d_{min}}$$

Problem

Let us design a singly reinforced concrete beam from the following data:

Clear span = 4 m

Width of supports = 300 mm

Service load = 5 kN/m

Materials: M20 Grade Concrete.

Fe-415 HYSD bars.

Solution:

1. Data:

Clear span = 4 m

Width of supports = 300 mm

Service load = 5 kN/m

Materials: M20 Grade Concrete.

2. Stresses:

$$f_{ck} = 20\ \text{N/mm}^2$$
$$f_y = 415\ \text{N/mm}^2$$

Load factor = 1.5 for dead and live loads.

3. Cross Sectional Dimensions:

Effective depth = (span/20) = (400/20) = 200 mm

Adopt d = 200 mm

D = 250 mm

b = 200 mm

Effective span = Clear span + effective depth

= (4 + 0.2) = 4.2 m

Centre of center of supports = (4 + 0.3) = 4.3 m

Hence L = 4.2 m

4. Loads:

$$\text{Self weight } = g = (0.2 \times 0.25 \times 25) = 1.25 \text{ kN/m}$$

$$\text{Live load } = q = 5.00 \text{ kN/m}$$

$$\text{Total load } = w = 6.25 \text{ kN/m}$$

$$\text{Design ultimate load } = w_u = (1.5 \times 6.25) = 9.375 \text{ kN/m}$$

5. Ultimate Moments and Shear Forces:

$$M_u = \left(0.125\, w_u L^2\right) = \left(0.125 \times 9.375 \times 4.2^2\right) = 20.67 \text{ kN.m}$$

$$V_u = \left(0.5\, w_u L\right) = \left(0.5 \times 9.375 \times 4.2\right) = 19.68 \text{ kN}$$

6. Tension Reinforcements:

$$M_u \lim = 0.138\, f_{ck}\, b d^2$$

$$= \left(0.138 \times 20 \times 200 \times 200^2\right) 10^{-6} \text{ kN.m}$$

$$= 22.08 \text{ kN.m}$$

Since $M_u < M_{u\,lim}$, section is under reinforced,

$$M_u = \left(0.87\, f_y\, A_{st}\, d\right)\left[1 - \frac{A_{st}\, f_y}{(b\, d\, f_{ck})}\right]$$

$$M_u = \left(0.87 \times 415\, A_{st} \times 200\right)\left[1 - \frac{415\, A_{st}}{(200 \times 200 \times 20)}\right]$$

Solving $A_{st} = 350 \text{ mm}^2$

Provide 3 bars of 16 mm diameter $\left(A_{st} = 402 \text{ mm}^2\right)$ and 2 hanger bars of 10 mm diameter.

7. Check for Shear Stress:

$$\tau_v = (V_u / bd) = (19.68 \times 10^3) / (200 \times 200) = 0.49 \text{ N/mm}^2$$
$$P_t = (100A_{st}) / (bd) = (100 \times 402) / (200 \times 200) = 1.005$$

$$\tau_c = 0.63 \text{ N/mm}^2 > \tau_v$$

Provide nominal shear reinforcements using 6 mm diameter two-legged stirrups at a spacing of:

$$s_v = \left(\frac{A_{st} 0.87 f_y}{0.4 \text{ b}}\right) = \left(\frac{2 \times 28 \times 0.87 \times 250}{0.4 \times 200}\right) = 152 \text{ mm}$$

But $s_v > 0.75 \text{ d } (> 0.75 \times 200) = 150 \text{ mm}$

Adopt spacing of stirrups as 150 mm centers.

8. Check for Deflection Control:

$$P_t = 1.005$$

$$(L/d)_{max} = (L/d)_{basic} \times K_t \times K_c \times K_f$$

$$= (20 \times 1.05 \times 1 \times 1)$$
$$= 21$$
$$(L/d)_{actual} = (4200/200) = 21$$

Since $(L/d)_{actual} = (L/d)_{max}$ deflection control is satisfactory.

9. Reinforcement Details:

The details of reinforcement in the beam is shown in the below figure:

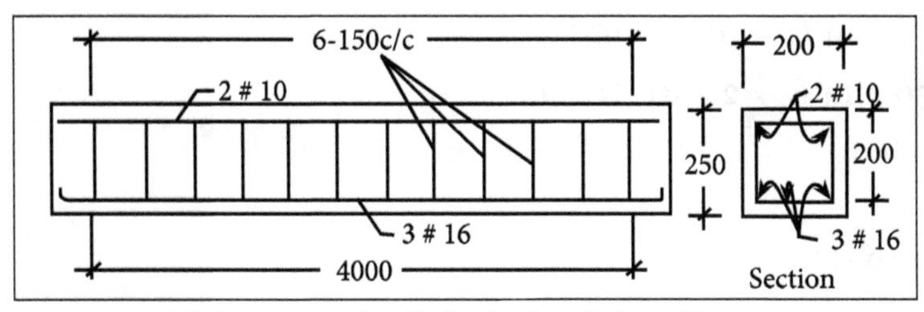

Reinforcement details in singly reinforced beam.

10. Design using SP-16 Design Tables:

$$(M_u / bd^2) = (20.67 \times 10^6) / (200 \times 200^2) = 2.58$$

$$A_{st} = (p_t\, b\, d)/100 = (1.005 \times 200 \times 200)/100 = 350\, mm^2$$

Hence A_{st} is the same as that computed using theoretical equations.

Doubly Reinforced Beams

A doubly reinforced concrete section is reinforced in both compression and tension regions. Then the section of the beam or slab can be a rectangle, T and L section. Therefore the necessity of using steel in the compression region arises because of two main reasons:

- When depth of the section is restricted, the strength that is available from a singly reinforced section is inadequate.
- At a support of a continuous beam or slab where bending moment changes sign. This situation can also arise in the design of a beam circular in plan.

Let us design a reinforced concrete beam of rectangular section:

Effective span = 5 m

Width of beam = 250 mm

Overall depth = 500 mm

Service load (DL + LL) = 40 kN/m

Effective cover = 50 mm

Materials: M-20 Grade concrete.

Solution:

1. Data:

$b = 250\ mm\ f_{ck} = 20\ N/mm^2$
$D = 500\ mm\ f_y = 415\ N/mm^2$
$d' = 450\ mm\ E_s = 2 \times 10^5\ N/mm^2$
$d' = 50\ mm$
$L - 5\ m$
$w = 40\ kN/m$

2. Ultimate Moments and Shear Forces:

$$M_u = (0.125 \times 1.5 \times 40 \times 5^2) = 187.5\ kN.m$$
$$V_u = (0.5 \times 1.5 \times 40 \times 5) = 150\ kN$$

3. Main Reinforcements:

$$M_{u\,lim} = 0.138 f_{ck}\, bd^2$$

$$= \left(0.138 \times 20 \times 250 \times 450^2\right) 10^{-6} \text{ kN.m}$$

$$= 140 \text{ kN.m}$$

Since $M_u > M_{u\,lim}$, design a doubly reinforced section.

$$\left(M_u - M_{u\,lim}\right) = (187.5 - 140) = 47.5 \text{ kN.m}$$

$$f_{sc} = \left\{\frac{0.0035\left(x_{u,\,max} - d'\right)}{x_{u,max}}\right\} E_s$$

$$= \left\{\frac{0.0035\left[(0.48 \times 450) - 50\right]}{(0.48 \times 450)}\right\} 2 \times 10^5$$

$$= 538 \text{ N/mm}^2$$

But $f_{sc} = 0.87, f_y = (0.87 \times 415) = 361 \text{ N/mm}^2$

$$\therefore A_{sc} = \left[\frac{\left(M_u - M_{u,lim}\right)}{f_{sc}(s - d')}\right]$$

$$= \left[\frac{47.5 \times 10^6}{361 \times 400}\right] = 329 \text{ mm}^2$$

Provide 2 bars of 16 mm diameter $\left(A_{sc} = 402 \text{ mm}^2\right)$,

$$A_{st2} = \left(\frac{A_{sc} f_{sc}}{0.87 f_y}\right) = \left(\frac{329 \times 361}{0.87 \times 415}\right) = 329 \text{ mm}^2$$

$$A_{st1} = \left[\frac{0.36\, f_{ck}\, b x_{u,lim}}{0.87\, f_y}\right]$$

$$= \left[\frac{(0.36 \times 20 \times 250 \times 0.48 \times 450)}{(0.87 \times 415)}\right] = 1077 \text{ mm}^2$$

Total tension reinforcement $= A_{st} = (A_{st1} + A_{st2})$

$$= (1077 + 329)$$
$$= 1406 \text{ mm}^2$$

Provide 3 bars of 25 mm diameter $(A_{st} = 1473 \text{ mm}^2)$

4. Shear Reinforcements:

$$\tau_v = (V_u / bd) = (150 \times 10^3)/(250 \times 450) = 1.33 \text{ N/mm}^2$$
$$p_t = (100\, A_{st})/(bd) = (100 \times 1473)/(250 \times 450) = 1.3$$

From IS: 456-2000

$$\tau_c = 0.68 \text{ N/mm}^2$$

Since $\tau_v > \tau_c$, shear reinforcements are required.

$$V_{us} = [V_u - (\tau_c\, b\, d)]$$
$$= 150 - (0.68 \times 250 \times 450)10^{-3}] = 73.5 \text{ kN}$$

Using 8 mm diameter 2 legged stirrups:

$$s_v = \left[\frac{0.87\, f_y\, A_{sv}\, d}{V_{us}}\right] = \left[\frac{0.87 \times 415 \times 2 \times 50 \times 450}{73.5 \times 10^3}\right]$$
$$= 221 \text{ mm}$$

$$s_v > 0.75\, d = (0.75 \times 450) = 337.5 \text{ mm}$$

Adopt a spacing of 200 mm near supports gradually increasing to 300 mm towards the centre of span.

5. Check for Deflection Control:

$$(L/d)_{actual} = (5000/450) = 11.1$$
$$(L/d)_{max} = [L/d)_{basic} \times K_t \times K_c \times K_f$$

$p_t = 1.3$ and $p_c = [(100 \times 402)/(250 \times 450)] = 0.35$

From figure $K_t = 0.93$(code book)

From figure $K_c = 1.10$

From figure $K_f = 1.00$

$$(L/d)_{max} = [(20 \times 0.93 \times 1.10 \times 1.00)] = 20.46$$

$$(L/d)_{actual} < (L/d)_{max}$$

Hence deflection control is satisfied.

6. Reinforcement Details:

The details of reinforcements in the doubly reinforced beam is shown in the below Figure:

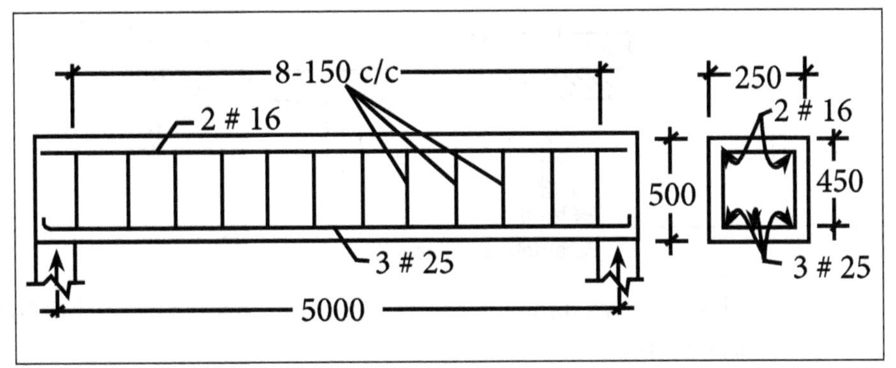

Reinforcement details in doubly reinforced beam.

7. Design using SP-16 Design Tables:

Compute the parameter:

$$\left(\frac{M_u}{b\,d^2}\right) = \left(\frac{187.5 \times 10^6}{250 \times 450^2}\right) = 3.7$$

$$p_t = 1.25 \text{ and } p_c = 0.310$$
$$A_{st} = (p_t bd)/100 = (1.25 \times 250 \times 450)/100 = 1406 \text{ mm}^2$$
$$A_{sc} = (p_c bd)1\,100 = (0.310 \times 250 \times 450)/100 = 348 \text{ mm}^2$$

The area of reinforcements are same as that computed using theoretical equations.

Flanged Beams

Design Parameters of Tee Beams

Most common type of reinforced concrete floor and roof system comprises of concrete slabs monolithically cast with floor beams in the span range of 5 to 10 m.

Here the compressive flange is made up of the width of rib and a portion of the slab length on either side of the rib referred to as the effective width of flange.

Below figure shows the prominent design parameters of flanged (Tee) beams.

Effective Width of Flange (b_f)

Effective width of flange should not be greater than the breadth of the web plus half the sum of the clear distances to the adjacent beams on either side.

For T-beams, $b_f = (L_o/6) + b_w + 6\ D_f$

For L-beams, $b_f = (L_o/12) + b_w + 3\ D_f$

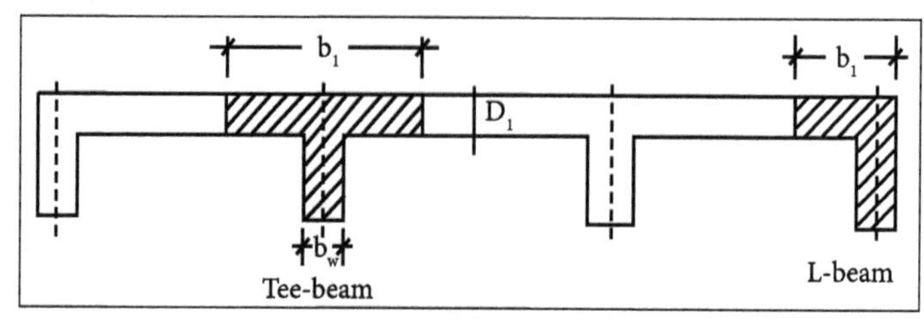

Parameters of flanged beams.

For isolated beams, the effective flange width shall be obtained as below but in no case greater than the actual width.

$$T-\text{beam},\ b_f = \left[\frac{L_o}{(L_o/b)+4} + b_w\right]$$

$$L-\text{beam},\ b_f = \left[\frac{L_o}{(L_o/b)+4} + b_w\right]$$

Where,

b_f = Effective width of flange.

L_o = Distance between points of zero moments in the beam (LQ = L for simply supported beams and 0.7

L for continuous beams).

L = Effective span.

b_w = Breadth of web.

D_f = Flange thickness.

b = Actual width of flange.

Effective Depth (d)

The basic span/depth ratios outlined are applicable for flanged beams modified by using the factor K_f which may also be termed as reduction factor.

For purpose of design, the span/depth ratios of the trial section is generally assumed in the range of 12 to 20 depending upon the span range and degree of loading as given in table.

Width of Web (b_w)

The web width of tee beam is influenced by the width of the supporting column or the width of the supports like brick concrete or stone masonry walls.

The nominal range of width of tee beams varies from 150 to 400 mm.

Flange Thickness (D_f)

The flange thickness is generally the same as the thickness of the slab between the ribs.

The slab thickness depends upon the spacing of ribs, type of loading and is governed by the basic span/depth ratios specified in table.

Thickness of the slab varies from a minimum of 100 mm to a maximum of 250 mm.

Reinforcement Requirements

The minimum percentage of reinforcement in a flanged beam is based on the width of web and effective depth.

The code specifies the minimum reinforcement as:

$$\left(\frac{A_s}{b_w d}\right) = \left(\frac{0.85}{f_y}\right)$$

The maximum percentage of tension reinforcement is flanged beams (based on rib width) is limited to 4 percent.

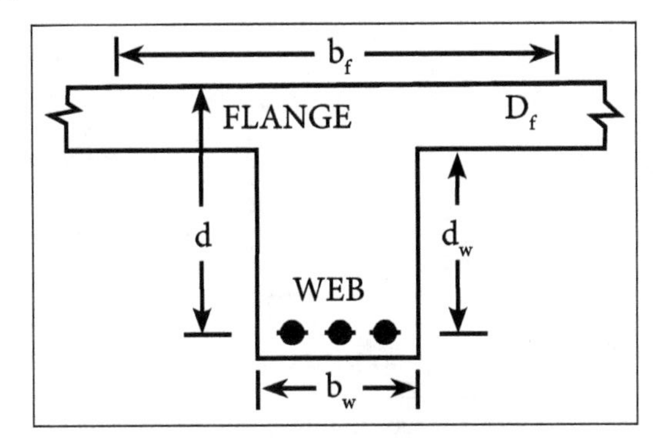

The figure shows the following four important dimensions of a T-beam section:

- Breadth of the flange (b_f).
- Thickness of the flange (D_f).
- Breadth of the rib (b_w).
- Depth of the rib (d_w).

Breadth of the Flange (b_f)

The breadth of the flange of the T-beam is that portion of the slab which acts monolithic with the beam and which resists the compressive stresses. According to Indian Standard Code (IS 456: 2000), a slab which is assumed to act as a flange of a T-beam shall satisfy the following:

- The slab be cast integrally with the web or the web and the slab shall be effectively bonded together in any other manner.
- If the main reinforcement of the slab is parallel to the beam, transverse reinforcement shall be provided. Such reinforcement shall not be less than 60% of the main reinforcement at the mid-span of the slab. The reinforcement is provided at the top of the slab.

Reinforcement in flanges of T-beams shall satisfy the above requirement (b). Where flanges are in tension, a part of the main tension reinforcement shall be distributed over the effective flange width or a width equal to one length of the span, whichever is smaller.

If the effective flange width exceeds one length of span, nominal longitudinal reinforcement shall be provided in the outer portion of flange. In absence of more accurate determinations, the effective width bf of the flange may be taken as the least of the following:

$$b_f = \frac{l_o}{6} + 6\,D_f + b_w \qquad \text{...(1)}$$

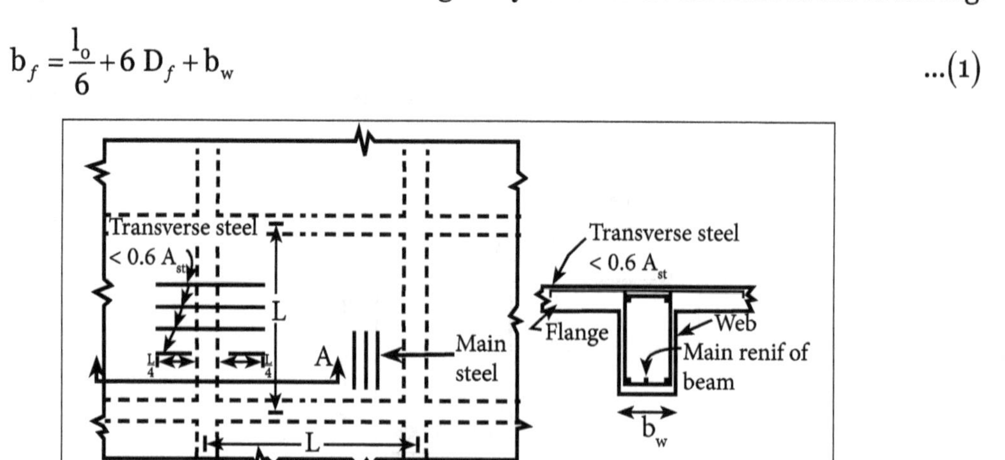

Provision of Transverse Steel for Flanged Section: (a) Plan (b) Section AA.

$$b_f = b_w + \frac{1}{2} \text{ (sum of clear distance to the adjacent sides on either beams)} \quad ...(2)$$

For isolated beams, the effective flange width shall be obtained as below but in no case greater than actual width:

$$b_f = \frac{l_o}{\left(\frac{l_o}{b}\right) + 4} + b_w \quad ...(3)$$

Where,

b_f = Effective width of flange.

l_o = Distance between points of zero moments in the beam.

b_w = Breadth of web.

D_f = Thickness of flange.

b = Actual width of flange.

For continuous beams and frames, lo may be assumed as 0.7 times the effective span.

Thickness of the Flange (D_f)

The thickness of the flange is taken equal to the total thickness of the slab, including cover. The slab spans in a direction transverse to the span of the beam and, therefore, the thickness of the slab is determined on the basis of its bending in transverse direction.

As far as design of the T-beam is concerned, D_f is the fixed dimension known from the considerations of bending etc., of the slab.

Breadth of the Rib or Web (b_w)

The breadth of the rib is equal to width of the portion of the beam in the tensile zone. This width should be sufficient to accommodate the tensile reinforcement with proper spacing between the bars.

The width of the beam should also be sufficient to provide lateral stability. From this point of view, it should atleast be equal to one-third of the height (depth) of the rib. From architectural point of view, the breadth of the rib should be equal to the width of the column supporting the beam.

Depth of Rib or Web (d_w)

The depth of rib is the vertical distance between the bottom of the flange and the centre of the tensile reinforcement and is dependent on the effective depth (d) of the beam. The effective depth (d) of the T-beam is the distance between the top of the flange and the centre of the tensile reinforcement. Thus, $d = d_w + D_f$.

The assumed overall depth of the T-beam may be taken as 1/12 to 1/15 of the span when it is simply supported at the ends. When it is continuous, the assumed overall depth may be taken as 1/15 to 1/20 of span for light loads, 1/12 to 1/15 of span for medium loads and 1/10 to 1/12 of span for heavy loads.

Design Procedure

The design of a T-beam is done in the following steps:

- Assume some suitable value of width of web based on criteria.
- Determine the effective width bf of the flange.
- For computation of self weight, assume total depth (D) of the beam, equal to 1/12 to 1/15 of the span when it is simply supported at the ends. When it is continuous, the assumed over all depth may be taken as 1/15 to 1/20 of the span for light loads, 1/12 to 1/15 for medium loads and 1/10 to 1/12 of span for heavy loads. Compute load on the beam.
- Assume effective span of the beam and compute. A ultimate design B.M. (M_{uD}) and ultimate S.F. (V_u).
- Compute effective depth of beam by taking 2/3 of the depth required for maximum sagging B.M., treating it as a balanced section. Thus,

$$d \, \underline{\Omega} \, \frac{2}{3}\sqrt{\frac{M_u}{R_u . b}}$$

Estimate total depth (D) of the beam as under:

$D = d +$ nominal cover + diameter (ϕ_s) of stirrups + $\frac{\varphi}{2}$ of estimated main bar.

Round off this value of D and recalculate d as under,

$d = D$ - nominal cover $- \phi_s - \phi/2$

Thus, effective depth d of the T-beam has been fixed.

- At this stage, b , b_f, D_f and d are known.

Assuming $x_u = D_f$, compute moment of resistance M_{u1}

$$M_{u1} = 0.36 f_{ck}\, b_f\, D_f \left(d - 0.416\, D_f\right)$$

- If $M_{u1} > M_{uD}$ the N.A. will fall inside the flange $\left(\text{i.e. } X_u \leq D_f\right)$. In that case, the reinforcement is given by,

$$M_u = M_{uD} = 0.87\, f_y\, A_{st}\, d\left[1 - \frac{f_y}{f_{ck}} \frac{A_{st}}{b_f . d}\right]$$

The solution of the above quadratic equation gives,

$$A_{st} = \frac{0.5\, f_{ck}}{f_y}\left[1 - \sqrt{1 - \frac{4.6\, M_{uD}}{f_{ck}\, b_f\, d^2}}\right] b_f\, .\, d$$

- If $M_{u1} < M_{uD}$, the N.A. falls outside the flange. Assuming $x_u = (7/3)\, D_f$ compute M_{u2}

$$M_{u2} = 0.36\, f_{ck} . b_w\left(\frac{7}{3} D_f\right)\left(d - 0.416 \times \frac{7}{3} D_f\right) + 0.446\, f_{ck}\left(b_f - b_w\right) D_f\left(d - 0.5\, D_f\right)$$

If $M_{u2} < M_{uD}, x_u > (7/3) D_f$, compute A_{sw} and A_{sf} as under,

$$A_{sw} = \frac{0.36 f_{ck}\, x_u\, b_w}{0.87 f_y} \quad \text{and} \quad A_{sf} = \frac{0.446\, f_{ck}\left(b_f - b_w\right) D_f}{0.87\, f_y}$$

Total $A_{st} = A_{sw} + A_{sf}$

When, $M_{uD} >> M_{u2}$ it is always advisable to compute $M_{u.lim}$. Since it is likely that $x_u > x_{u.Max}$. If $M_{uD} > M_{u.lim}$ the section will be doubly reinforced.

- If $M_{u2} > M_{uD}$, $x_u < (7/3)\, D_f$. The design procedure is this case will be that of trial and error, with the following steps:
  - Assume $x_u < (7/3) D_f$.
  - Compute $y_f = 0.15\, x_u + 0.65\, D_f$ (subject to a max. of D_f).
  - Compute M_u,
  - $M_u = 0.36 \frac{x_u}{d}\left(1 - 0.416 \frac{x_u}{d}\right) f_{ck}\, b_w\, d^2 + 0.446\, f_{ck}\left(b_f - b_w\right) y_f\left(d - 0.5\, y_f\right)$
  - Compare M_u with M_{uD},

If $M_u = M_{uD}$, assumed value of X_u is correct.

If $M_{uD} > M_u$, increase x_u in next trial.

If $M_{uD} < M_u$, decrease x_u in next trial.

Repeat till $M_u = M_{uD}$.

- Knowing x_u, compute $C_u = 0.36 f_{ck} x_u b + 0.446 f_{ck} (b_f - b_w) y_f$ and $A_{st} = \frac{C_u}{0.87 f_y}$.

Alternatively, x_u can be found by substituting the value of y_f in terms of x_u and then solving the quadratic equation in terms of x_u. This value of x_u will be acceptable only if the value of y_f obtained from this is less than (or equal to) D_f.

- Check for shear and design the shear reinforcement.
- Check for anchorage and development length at the supports.

Problem

1. The floor of a hall measures 16 m × 6 into the faces of the supporting walls. The floor consists of three beams spaced at 4 m centre to centre and the slab thickness is 120 mm. The floor carries a uniformly distributed load of 5 kN/m², inclusive of the floor finishes. Let us design the intermediate beam. Use M20 concrete and Fe415 steel. The support width may be assumed equal to 500 mm.

Solution:

Given:

Three beams spaced = 4 m centre to centre

Slab thickness = 120 mm

Uniformly distributed load $= 5 \text{ kN/m}^2$

Step 1: Determination of b_w.

The width of the web should be sufficient to accommodate the tensile reinforcement. Assume that the reinforcement consists of 5 bars of 25 mm diameter and the stirrups consists of 8 mm φ wire.

Keeping a nominal cover of 25 mm at the sides and 25 mm clear distance between the bars, total width $b_w = (5 \times 25) + (4 \times 25) + (2 \times 25) + (2 \times 8) = 291 \text{ mm}$. Keep $b_w = 300$ mm.

Step 2: Computation of width of flange b_f.

The width of flange is taken the least of the following:

$$b_f = \frac{l_o}{6} + b_w + 6 D_f \text{ (where } l_o = 6 + 0.5 = 6.5 = 6500 \text{ mm)}$$

$$= \frac{6500}{6} + 300 + 6 \times 120 \,\Omega\, 2100 \text{ mm}$$

or

$$b_f = b_w + \frac{1}{2}(\text{sum of clear distances to adjacent beams on either side}).$$

$$= 300 + \frac{1}{2}\left[2(4000 - 300)\right] = 4000 \text{ mm.}$$

Hence adopt $b_f = 2100$ mm.

Step 3:

Effective span = 6 + 0.5 = 6.5 m = 6500 mm

Assume $D = \frac{1}{13} 6500$ of span $= \frac{6500}{13} = 500$ mm (say)

Dead load of slab / m² = 0.12 × 1 × 1 × 25000 = 3000 N/m²

Super-imposed load / m² = 5000 N/m²

Total = 8000 N / m²

Load per metre run of beam = 8000 × 4 = 32000 N/m

Self weight of web/m run = 0.3 × 0.38 × 1 × 25000 = 2850 N/m

Total load w = 34850 Ω 35000 N/m (say)

$$w_u = 1.5,\ w = 1.5 \times 35000 = 52500 \text{ N/m}$$

Step 4: Computation of M_u and V_u.

$$M_u = \frac{w_u L^2}{8} = \frac{52500(6.5)^2}{8} \,\Omega\, 277265 \text{ N-m} = 277.265 \times 10^6 \text{ N-mm}$$

$$V_u = \frac{w_u l}{2} = 52500 \times \frac{6}{2} = 157500 \text{ N}$$

Step 5: Fixation of Effective Depth and Total Depth.

$$d = \frac{2}{3}\sqrt{\frac{M_u}{R_u . b}} = \frac{2}{3}\sqrt{\frac{277.265 \times 10^6}{2.761 \times 300}} = 386$$

Also, from stiffness point of view, L/d = 20.

$p_{lim} = 0.955\,\%\ \Omega\ 1\%$, for which, modification factor F_t = I.

Also, $b_w / b_f = \frac{300}{2100} \underline{\Omega}\ 0.14$ for which reduction factor $F_b = 0.8$.

Hence $d = \frac{L}{20 \times 1 \times 0.8} = \frac{6500}{20 \times 1 \times 0.8} = 406\ \text{mm}$

Adopting d = 406 mm (greater of the two), using 25 mm φ main bars, 8 mm φ stirrups and providing 25 mm nominal cover, D = 406 + 25 + 8 + 12.5 = 451.5 mm.

However, keep D = 450 mm, so that d = 450 - 25 - 8 - 25/2 = 404.5 mm.

Thus, for the T-beam, we have $b_w = 300$ mm, $b_f = 2100$ mm; $D = 450$ mm and $d = 404.5$ mm.

Step 6:

Assume $x_u = D_f = 120$ mm

$$M_{u1} = 0.36\, f_{ck}\, b_f\, D_f\,(d - 0.416\, D_f) = 0.36 \times 20 \times 2100 \times 120\ (404.5 - 0.416 \times 120)$$

$$= 643.35 \times 10^6\ \text{N-mm}$$

Step 7:

Since $M_{u1} > M_{uD}$, N.A. falls inside the flange. i.e. $x_u \le D_f$

Hence $M_u = M_{uD} = 0.87\, f_y\, A_{st}\, d \left[1 - \frac{f_y\, A_{st}}{f_{ck}\, b_f d}\right]$

$$A_{st} = \frac{0.5\, f_{ck}}{f_y}\left[1 - \sqrt{1 - \frac{4.6\, M_{uD}}{f_{ck}\, b_f\, d^2}}\right] b_f\, d$$

$$= \frac{0.5 \times 20}{415}\left[1 - \sqrt{1 - \frac{4.6 \times 277.265 \times 10^6}{20 \times 2100 (404.5)^2}}\right] \times 2100 \times 404.5$$

$$= 1997\ \text{mm}^2$$

Choosing 25 mm φ bars, number of bars = 1997/490.9 = 4.07.

Provide 5-25 mm φ bars. Actual A_{st} = 5 ×490.9 = 2454.5 mm^2.

Bend 2 bars up at 45° at a distance of 1.414 (0.9 d) = 1.414 × 0.9 × 404.5 ≏ 500 mm from the edge of the support.

Step 8:

$$\tau_v = \frac{V_u}{b_w d} = \frac{157500}{300 \times 404.5} \underline{\Omega}\ 1.3\ \text{N/mm}^2$$

$$\frac{100\ A_s}{b_w d} = \frac{100 \times 3 \times 490.9}{300 \times 404.5} \Omega\ 1\%$$

For 1 % reinforcement, $\tau_c = 0.62\ N/mm^2$; Also $\tau_{c\ .\ max} = 2.8\ N/mm^2$. Hence safe.

Since $\tau_v > \tau_c$, shear reinforcement is necessary,

$$V_{uc} = \tau_c\, b_w .\, d = 0.62 \times 300 \times 404.5 = 75236\ N$$

$$V_{us} = 157500 - 75236 - 82264\ N$$

Shear resistance of 2 bent up bars, is given by,

$$V_{us1} = 0.87\ f_y\ A_{sv}\ \sin 45^\circ = 0.707 \times 0.87 \times 415\ (2 \times 490.9) = 250616\ N$$

However, maximum resistance assigned to the bent-up bars $-\ _{us}_$41132 N.

Hence provide vertical stirrups for the remaining 41132 N shear $= V_{us2}$.

Using 8 mm φ 2-legged stirrups, $A_{sv} = 2\frac{\pi}{4}(8)^2 = 100.5\ mm^2$

$$S_v = \frac{0.87\ f_y\ A_{sv}\ d}{V_{us2}} = \frac{0.87 \times 415 \times 100.5 \times 404.5}{41132} = 356.8\ mm$$

Maximum spacing corresponding to nominal stirrups is given by,

$$S_v = \frac{2.175\ A_{sv}\ f_y}{b_w} = \frac{2.175 \times 100.5 \times 415}{300} = 302\ mm$$

Also, the spacing should be less than lesser of 0.75 d (= 0.75 × 404.5 = 303) or 300 mm.

Hence provide 8 mm φ 2 Iegged stirrups @ 300 mm c/c throughout. Provide 2-12 mm φ bars top as anchor bars.

Step 9: Check for Development Length at Supports.

The code envisages that the diameter of bars should be so selected that following relation is satisfied,

$$1.3\frac{M_{1u}}{V_u} + L_o \geq L_d.$$

A_{st1} = Area of remaining bars at supports

$$= 3 \times 490.9 = 1472.7\ mm^2$$

$$x_u = \frac{0.87\ f_y\ A_{st1}}{0.36\ f_{ck}\ b_w} = \frac{0.87 \times 415 \times 1472.7}{0.36 \times 20 \times 300} = 246.2 \text{ mm}$$

$$M_{1u} = 0.87\ {}_yA_{st1}\left(d - 0.146\ x_u\right)$$

$$= 0.87 \times 415 \times 1472.7\ (404.5 - 0.416 \times 246.2)$$
$$= 160.62 \times 10^6 \text{ N-mm}$$

$$V_u = \frac{w_u L}{2} = 52500 \times \frac{6.5}{2} = 170625 \text{ N}$$

$$L_d = 47\ \phi = 47 \times 25 = 1175 \text{ mm}$$

Taking the bars straight into supports without any hook or bend and providing side cover x' = 30 mm we have,

$$L_o = \frac{L_s}{2} - x' = \frac{500}{2} - 30 = 220 \text{ mm}$$

$$1.3\ \frac{M_{1u}}{V_u} + L_o = 1.3\ \frac{160.62 \times 10^6}{170625} + 220 = 1223 + 220 = 1443 \text{ mm} > L_d$$

Hence Code requirements are satisfied.

Step 10: Details of Reinforcement.

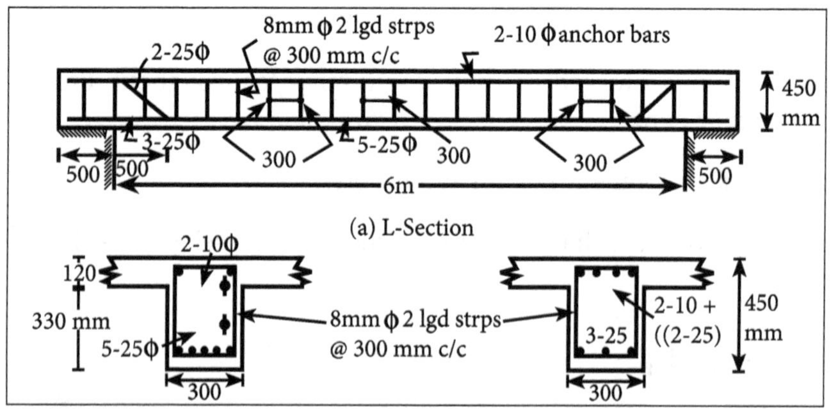

(a) Section at mid span (b) Section at supports Reinforcement detail.

Design Parameters of L-beams

Edge beams which are cast monolithic with slabs on one side of the rib are designated as L-beams. Torsional moments develops in the beams in addition to the bending moments and shear forces. The torsional and hogging bending moments are maximum at the support sections.

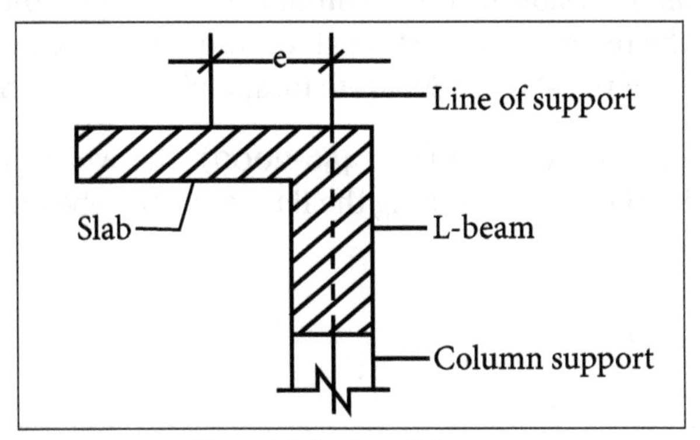

Eccentric load on L-beam.

Support section of the L-beam is the most critical section subjected to combined bending, torsion and shear. This section is designed according to the provisions of the IS: 456 code.

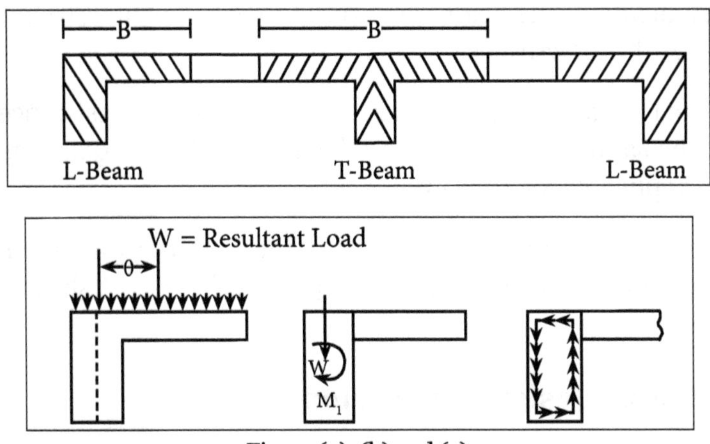

Figure (a), (b) and (c).

In the case of T-beam, the rectangular beam has flanges on both the sides and hence load is transferred to it in vertical plane passing through the middle of its width. However in the case of monolithic beam-slab construction, the end beams have flanges to one side only, as in figure (a). Such a beam having the shape of letter 'L' is known as L-beam.

Figure (b) shows the loading on the beam received from the slab. Evidently, the beam is not symmetrically loaded. Due to this, an eccentric load W is transferred to the beam. This eccentric load may be considered as central load W plus a moment M_1 as shown in figure (c).

The central load W causes bending moment in the beam, which is jointly resisted by the rectangular portion of the beam as well as the flange (of width B), similar to that of a T-beam.

The moment M_1 causes torsion in the beam and is known as the torsional moment which is resisted by the rectangular portion alone and the flange does not contribute to any torsional moment of resistance. Separate torsional resistance has to be provided.

According to Indian Standard Code (IS : 456-2000), the width B (or b_f) of the slab, acting monolithic with the beam, forming the flange of the L-beam is taken as the least following:

$$b_f = \frac{l_0}{12} + b_w + 3D_f \qquad ...(1)$$

$$b_f = \frac{0.5\, l_0}{\frac{l_0}{b} + 4} + b_w \qquad ...(2)$$

(but in no case greater than the actual width of flange) where l_0 is the distance between points of zero moments in the beam. For continuous beams and frames l_0 may be assumed as 0.7 times the effective span.

The location of the N.A. and the calculations of the moment of resistance of L-beam are done exactly in the same way as that for T-beam and all the equations developed for T-beam are also applicable for L-beam.

The design for bending moment is also done exactly in the way as that for T-beam. However, additional reinforcement has to be provided to resist shear induced by torsional moment.

Design of L-Beam: Design for Torsion

The design principles for L-beam are similar to those for T-beam, except that additional torsional reinforcement is provided to resist torsional moment. The width b_f of the slab, acting monolithic with the beam, forming the flange of the L-beam is taken as the least of the following:

- $b_f = \frac{l_0}{12} + b_w + 3\,D_f$
- For isolated L-beam, $b_f = \frac{0.5\, l_0}{\frac{l_0}{b} + 4} + b_w$

But in no case greater than the actual width of the flange, where l_0 is the distance between points of zero moments in the beam. For continuous beams and frames, l_0 may be assumed to be 0.7 times the effective span.

The location of N.A. and calculations of the moment of resistance of L-beam are done exactly in the same way as that for a T-beam and all the equations developed for T-beam are also applicable for L-beam.

Design for Shear and Torsion

The design for shear and torsion is done in the following steps:

- Calculate shear force V_u and torsional moment T_u.
- Compute the equivalent shear stress V_e and τ_{ve} from the expressions.

$$V_e = V_u + 1.6\frac{T_u}{b} \text{ and } \tau_{ve} = \frac{V_e}{b_w.d}$$

The value of τ_{ve} should not exceed τ_c. If it exceeds, the section should be redesigned by increasing the concrete.

- If τ_{ve} does not exceeds τ_c, provide nominal shear stirrups given by the equation, $s_v \le \frac{2.175\, A_{sv}\, f_y}{b_w}$.
- If τ_{ve} exceeds τ_c provide shear and torsion reinforcement in the form of longitudinal as well as transverse reinforcement in steps below.
- Longitudinal reinforcement: It is designed to resist an equivalent B.M. M_{e1} given by $M_{e1} = M_u + M_T$ where M_u = B.M. at the cross-section and $M_T = T_u\left(\frac{1+D/b_w}{1.7}\right)$.

If M_T exceeds M_u, provide longitudinal reinforcement on the compression face also for a moment M_{e2} given by $M_{e2} = M_T - M_u$, the moment M_{e2} being taken as acting in the opposite sense to the moment M_u.

- Transverse reinforcement: Find the area of transverse reinforcement A_{sv} given by,

$$A_{sv} = \frac{T_u\, s_v}{b_1 d_1 (0.87\, f_y)} + \frac{V_u\, s_v}{2.5\, d_1 (0.87\, f_y)}, \text{ subject to a minimum of } \frac{(\tau_{ve} - \tau_c) b_w\, s_v}{0.87\, f_y}$$

Where,

b_1 = Centre to centre distance between corner bars in the direction of the width.

d_1 = Centre to centre between corner bars, in the direction of depth.

The spacing s_v should not exceed the least of x_1, $(x_1 + y_1)/4$ and 300 mm, where x_1 and y_1 are respectively the short and long dimensions of the stirrup.

- Side face reinforcement: Where the depth of the web in a beam exceeds 450 mm, side face reinforcement shall be provided along the two faces. The total area of such reinforcement shall be not less than 0.1 percent of the web area and shall be distributed equally on the two faces at spacing not exceeding 300 mm or web thickness whichever is less.

Figure shows the plan and section of a canopy to be provided over an entrance opening of a building. Let us design the beam and the slab. Use M 20 concrete and Fe 415 steel, the super-imposed load on the slab may be taken as 1200 N/ m².

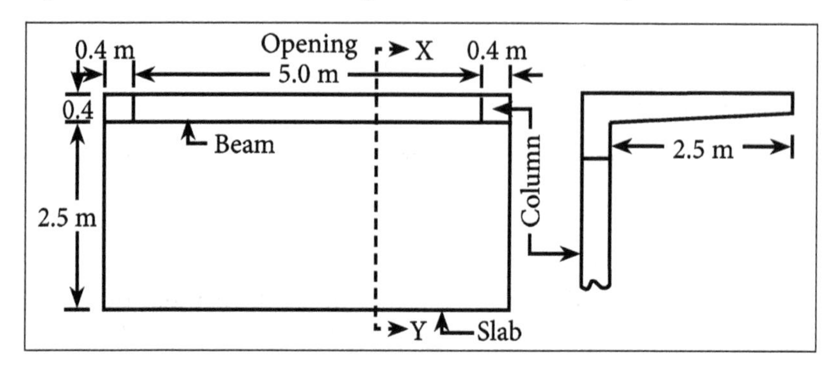

Solution:

Given:

M 20 concrete and Fe 415 steel.

Super-imposed load = 1200 N/ m².

Since the beam and the slab are monolithic, the beam will act as L-beam and part of the slab will resist the compression of the beam. The breadth of the beam will be kept equal to the width of the column, i.e. 400 mm.

Design of the Canopy Slab

Computation of Loading, B.M. and S.F.

Let the average depth the slab be 120 mm, for the purpose of computing dead weight. Dead load due to self weight $= 0.12 \times 1 \times 1 \times 25000 = 3000$ N/m².

Super-imposed load $=$ 1200 N/m²

$$M = \frac{4200(2.5)^2}{2} \times 1000 = 13.125 \times 10^6 \text{ N-mm}$$

Total $w = 3000 + 1200 = 4200 \text{ N/m}^2$

$$M_u = 1.5 \times 13.125 \times 106 \ \underline{\Omega} \ 19.69 \times 10^6 \text{ N-mm}$$

$$V = 4200 \times 2.5 = 10500 \text{ N}; \ V_u = 1.5 \times 10500 = 15750 \text{ N}$$

Computation of Effective Depth and Total Depth

For Fe 415 Steel, $\frac{x_{u.max}}{d} = \frac{700}{1100 + 0.87 \times 415} = 0.479$

$$R_u = 0.36\, f_{ck} \frac{x_{u.max}}{d}\left(1 - 0.416 \frac{x_{u.max}}{d}\right) = 0.36 \times 20 \times 0.479\,(1 - 0.416 \times 0.479) = 2.761$$

$$d = \sqrt{\frac{M_u}{R_u\, b}} = \sqrt{\frac{19.69 \times 10^6}{2.761 \times 1000}} = 84.4 \text{ mm}$$

However, for stiffness, the ratio of span to depth of cantilever = 7. For an under-reinforced slab, assuming p_t = 0.2 %, we get a modification factor of 1.7 for Fe 415 steel. Also, assuming d = 200 mm, effective span L = 2.5 + 0.2/2 = 2.6 m.

Hence, $d = \frac{2600}{7 \times 1.7} = 219$ mm.

Hence keep D = 250 mm at the supports and reduce it to 100 mm at the free end. Using 15 mm nominal cover and 8 mm φ bars, available d = 250 - 15 - 4 = 231 mm at the fixed end and d = 100 - 15 - 4 = 81 mm at the free end.

Average total thickness = (250 + 100)/2 = 175 mm against assumed value of 120 mm.

Hence dead load due to revised dimension = 0.175 × 25000 = 4375 N.

Thus, total w = 4375 + 1200 = 5575 N/m².

Revised, $M = \frac{5575(2.5)^2}{2} \times 1000 = 17.422 \times 10^6$ N-mm

and $M_u = 1.5 \times 17.422 \times 10^6 = 26.132 \times 10^6$ N-mm

Revised, $V = 5575 \times 2.5 = 13938$ N and $V_u = 1.5 \times 13938 = 20906$ N.

Computation of Steel Reinforcement

For an under-reinforced section,

$$A_{st} = \frac{0.5\, f_{ck}}{f_y}\left[1 - \sqrt{1 - \frac{4.6\, M_u}{f_{ck}\, bd^2}}\right] bd$$

$$= \frac{0.5 \times 20}{415}\left[1-\sqrt{1-\frac{4.6 \times 26.132 \times 10^6}{20 \times 1000(231)^2}}\right] \times 1000 \times 231 = 322.8 \text{ mm}^2$$

Spacing of 8 mm φ bars, is $s = \frac{1000 \times 50.26}{322.8} = 155.7$ mm

Hence provide 8 mm φ @ 150 mm c/c.

Available $= A_{st} = \frac{1000 \times 50.26}{150} = 335 \text{ mm}^2$; $p_t = \frac{335 \times 100}{1000 \times 231} = 0.145$ %

Computation of Distribution Reinforcement

$$A_{sd} = \frac{0.12}{100} \times \text{breadth} \times \text{average depth} = \frac{0.12}{100} \times 1000 \times 175 = 210 \text{ mm}^2$$

Spacing of 8 mm ϕ bars $= 1000 \times \frac{50.27}{210} = 218$ mm.

However provide 8 mm ϕ bars 210 mm c/c.

Check for Development Length at the Support

At the support, $A_{st1} = 335 \text{ mm}^2$

$$x_u = \frac{0.87\, f_y\, A_{st1}}{0.36\, f_{ck}\, b} = \frac{0.87 \times 415 \times 335}{0.36 \times 20 \times 1000} = 16.8 \text{ mm}$$

$$M_{1u} = 0.87 f_y\, A_{st1}\left(d - 0.416\, x_u\right) = 0.87 \times 415 \times 335\ (231 - 0.416 \times 16.8)$$
$$= 27.09 \times 10^6 \text{ N-mm}$$

$V_u = 15750$ N ; $L_d = 47\,\phi = 47 \times 8 = 376$ mm.

Taking the bars straight into supports, without any hook or bend and providing side cover x' = 20 mm, we have $L_o = 100/2 - 20 = 30$ mm.

$$\therefore 1.3\,\frac{M_{1u}}{V_u} + L_o = 1.3 \times \frac{27.09 \times 10^6}{15750} + 30 = 2266 \text{ mm} > L_d.$$

Hence OK. 1575.

Design of L-Beam

Fixation of b_w, b_f and D

Let $b_w = 400$ mm; Effective span $l_o = 5 + 0.4 = 5.4$ m

For an isolated beam, b = 400 + 2500 = 2900 mm

$$b_f = \frac{0.5\, l_o}{\left(\frac{l_o}{b}\right) + 4} + b_w = \frac{0.5 \times 5400}{\frac{5400}{2900} + 4} + 400 \underline{\Omega} 860 \text{ mm}$$

Thickness of slab at (0.86 - 0.4) = 0.46 m from the face of the beam,

$$= 100 + \frac{250 - 100}{2.5}(2.5 - 0.46) = 222.4 \text{ mm}$$

$$\therefore \text{Average } D_f = \frac{250 + 222.4}{2} \underline{\Omega}\ 236 \text{ mm}$$

Assume over all depth =1/13 = 6000/13 ≈ 460 mm for the purpose of computation of dead load.

Computation of Load and B.M.

Load from the slab $= 2.5 \times 1 \times 5575 \underline{\Omega}\ 13938$ N/m

Dead load of beam /m = 0.46 × 0.4 × l × 25000 = 4600 N/m

Total $w = 13938 + 4600 \underline{\Omega}\ 18540$ N/m.

$$\text{Assuming partial fixity, } M = \frac{wL^2}{10} = \frac{18540(5.4)^2}{10} = 54060 \text{ N-m}$$

$$\therefore M_u = 1.5 \times 54050 \times 10^3 = 81.09 \times 10^6 \text{ N-mm}$$

Also, $V = wL = 18540 \times 2.5 = 46350$; $V_u = 1.5 \times 46350 = 69525$ N

Fixation of Effective Depth and Total Depth

$$\text{Let } d\ \underline{\Omega}\ \frac{2}{3}\sqrt{\frac{M_u}{R_u\, b_w}} = \frac{2}{3}\sqrt{\frac{81.09 \times 10^6}{2.761 \times 400}} = 271 \text{ mm}$$

Also, from stiffness point of view L/d = 20. Taking $p_t = p_{t.Lim}$ = 1%. We get modification factor $F_t = 1$.

Also, $b_w / b_f = 400/860 = 0.465$, for which reduction factor $F_b = 0.85$. Hence $d = 5400/(20 \times 1 \times 0.85) = 318$ mm.

Depth from shear point of view: The total thickness should be such that the beam is safe in shear due to both bending as well as torsion.

$$T = \left[5575 \times 2.5\left(\frac{2.5}{2} + \frac{0.4}{2}\right)\right]\frac{5.4}{2} \text{ N-m} = 54565 \text{ N-m} = 54.565 \times 10^6 \text{ N-m}$$

$$V = \frac{wl}{2} = \frac{18540(5)}{2} = 46350 \text{ N}$$

$$T_u = 1.5 \times 54.565 \times 10^6 = 81.848 \times 10^6 \text{ N-mm and } V_u = 1.5 \times 46350 = 69525 \text{ N}$$

$$V_e = V_u + 1.6(T_u / b_w) = 69525 + 1.6 \times \frac{81.848 \times 10^6}{400} = 396915 \text{ N}$$

$$\tau_{ve} = V_e / b_w d.$$

Limiting this to $\tau_{c,max} = 2.8$ N/mm² for m 20 concrete, we get,

$$d = \frac{V_e}{b_w \ \tau_{c\,max}} = 396915/(400 \times 2.8) = 354 \text{ mm.}$$

However, keep D = 450 mm, so that providing 25 mm nominal cover and using 16 mm ϕ main bars and 8 mm diameter stirrups, available d = 450 - 25 = 8 - 16/2 = 409 mm.

Determination of Longitudinal Reinforcement

$$M_{e1} = M_u + M_T, \text{ where } M_u = 81.09 \times 10^6 \text{ N-mm}$$

$$M_T = T_u\left(\frac{1 + D/b_w}{1.7}\right) = 81.848 \times 10^6\left[\frac{1 + 450/400}{1.7}\right] = 102.31 \times 10^6 \text{ N-mm}$$

$$M_{e1} = (81.09 + 102.31)10^6 = 183.4 \times 10^6 \text{ N-mm}$$

Let us find the longitudinal reinforcement for this equivalent B.M.

For Fe 415 - M 20 combination, $R_u = 2.761$

$$\therefore M_{u\,.\,lim} = R_u b_w d^2 + 0.446 f_{ck}(b_f - b_w) y_f (d - 0.5\, y_f)$$

where $y_f = 0.15\, x_{u\,.\,max} + 0.65\, D_f = 0.15 \times 0.479 \times 409 + 0.65 \times 236 = 182.8 \text{ mm} < D_f$

$$M_{u\,.\,lim} = 2.761 \times 400\,(409)^2 + 0.446 \times 20(860-400) \times 182.8\,(409-0.5 \times 182.8)$$

$$= 423 \times 10^6 \text{ N-mm}$$

Since $M_{e1} < M_{u\,.\,lim}$ = the section is under-reinforced, for which,

$$A_{st1} = \frac{0.5\,f_{ck}}{f_y}\left[1-\sqrt{1-\frac{4.6\,M_{e1}}{f_{ck}\,b_f\,d^2}}\right]b_f\,d$$

$$= \frac{0.5 \times 20}{415}\left[1-\sqrt{1-\frac{4.6 \times 183.4 \times 10^6}{20 \times 860(409)^2}}\right] \times 860 \times 409$$

$$= 1350 \text{ mm}^2$$

Using 16 mm φ bars, number of bars = 1350/2017.

Hence provide 7 bars of 16 mm φ at the bottom of the beam, throughout its length.

Also, $M_{e2} = M_T - M_U = 102.31 \times 10^6 - 81.09 \times 10^6 = 21.22 \times 10^6$ N-mm.

Since the section is under-reinforced for M_{e2}, the area of steel is given by,

$$A_{st2} = \frac{0.5\,f_{ck}}{f_y}\left[1-\sqrt{1-\frac{4.6\,M_{e2}}{f_{ck}\,b_w\,d^2}}\right]b_w\,d$$

$$= \frac{0.5 \times 20}{415}\left[1-\sqrt{1-\frac{4.6 \times 21.22 \times 10^6}{20 \times 400(409)^2}}\right] \times 400 \times 409$$

$$= 146 \text{ mm}^2$$

This is quite small. However provide 3 bars of 16 mm ϕ to which will also work as holding bars for shear stirrups.

Determination of Transverse Reinforcement

Transverse reinforcement will be provided in the form of vertical stirps. Let the top 16 mm ϕ corner bars be provided at a nominal cover of 25 mm. Let us use 4-legged stirrups of 10 mm φ bar, having $A_{sv} = 4 \times \frac{\pi}{4}(10)^2 = 314 \text{ mm}^2$.

b_1 = Centre to centre distance between corner bars in the direction of width = 400 - 2 (25 + 10 + 8) = 314 mm.

d_1 = Centre to centre distance between corner bars in the direction of depth = 450 - 2 (25 + 10 + 8) = 364 mm.

Now, $A_{sv} = \frac{T_u\, s_v}{b_1\, d_1 (0.87\, f_y)} + \frac{V_u \cdot s_v}{2.5\, d_1 (0.87\, f_y)}$

$$\therefore 314 = \left(\frac{81.848 \times 10^6}{314 \times 364\,(0.87 \times 415)} + \frac{69525}{2.5 \times 364 \times 0.87 \times 415} \right) s_v.$$

From which, $s_v = 143$ mm,

As per code requirements, this spacing should not exceed x_1, $(x_1 + y_1)/4$ and 300 mm

Where,

x_1 = Short dimension of stirrup = 400 - 2 (16 + 5) = 358 mm

Y_1 = Long dimension of stirrup = 450 - 2 (16 + 5) = 408 mm

$(x_1 + y_1)/4 = (358 + 408)/4 = 191.5$ mm

Hence provide 10 mm ϕ 4 legged stirrups @ 140 mm c/c at supports.

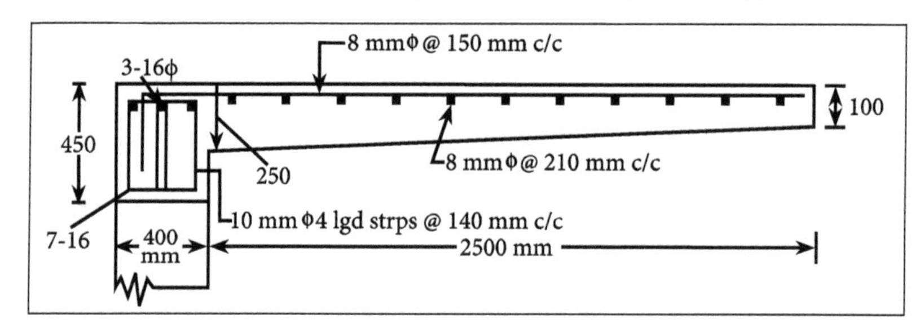

However, at the mid-span of the beam. where both V_u and T_u are zero, provide 8 mm ϕ 4 Iegged stirrups @ 300 mm c/c for the middle 2.5 m length of the beam. The details of reinforcement are shown in the figure.

Problem

1. Let us consider a tee beam slab floor of an office comprises of a slab 150 mm thick spanning between ribs spaced at 3 m centers. The effective span of the beam is 8 m. Live load on floor is 4 kN/m². Using M-20 grade concrete and Fe-415 HYSD bars, Design one of the intermediate tee beams.

Solution:

Given:

L = 8 m

$D_f = 150$ mm

$q = 4\ kN/m^2$

Spacing of tee beams = 3 m

$f_{ck} = 20\ N/mm^2$

$f_y = 415\ N/mm^2$

1. Cross Sectional Dimensions:

Basic span/depth ratio for simply supported beams is 20.

For tee beams, assuming the width of rib = 300 mm and flange width = 3 m, the ratio of web width to flange width is equal to (300/3000) = 0.1.

Reduction factor = 0.8

Hence basic span/depth ratio = (20 × 0.8) = 16

$\therefore\ d-(span/16)=(8000/16)=500\ mm$

Adopt overall depth = D = 550 mm with cover = 50 mm.

Hence the tee beam parameters are:

$d = 500\ mm$

$D = 550\ mm$

$b_w = 300\ mm$

$D_f = 150\ mm$

2. Loads:

Self weight of slab = (0.15 × 25 × 3) = 11.25 kN/m

Floor finish = (0.6 × 3) = 1.80

Self weight of rib = (0.3 × 0.4 × 25) = 3.00

Plaster finishes = 0.45

Total dead load = g = 16.50 kN/m

Live load = q = 4.00 kN/m

Design ultimate load $= w_u = 1.5\ (16.50+4.0) = 30.75\ kN/m$

3. Ultimate Moments and Shear Forces:

$$M_u = (0.125 \times 30.75 \times 8^2) = 246 \text{ kN.m}$$

$$V_u = (0.5 \times 30.75 \times 8) = 123 \text{ kN}$$

4. Effective Width of Flange:

$$b_f = [(L_o/6) + b_w + 6D_f]$$
$$= [(8/6) + 0.3 + (6 \times 0.15)]$$
$$= 2.53 \text{ mm}$$
$$= 2530 \text{ mm}$$

Centre to center of ribs = (3 - 0.3) = 2.7 m

Hence the least of (i) and (ii) is $b_f = 2530$ mm.

5. Moment Capacity of Flange:

$$M_{uf} = 0.36\, f_{ck} b_f D_f (d - 0.42\, D_f)$$

$$= 0.36 \times 20 \times 2530 \times 150\ (500 - 0.42 \times 150)$$
$$= 1194 \times 10^6 \text{ N.mm}$$
$$= 1194 \text{ kN.m}$$

Since $M_u < M_{uf}$, $x_u < D_f$

Hence the section is considered as rectangular with b_f.

6. Reinforcements:

$$M_u = 0.87\, f_y\, A_{st}\, d \left[1 - \left(\frac{A_{st} f_y}{b\, d\, f_{ck}}\right)\right]$$

$$(246 \times 10^6) = (0.87 \times 415 \times A_{st} \times 500)\left[1 - \frac{415\, A_{st}}{(2530 \times 500 \times 20)}\right]$$

Solving $A_{st} = 1417 \text{ mm}^2$

Provide 3 bars of 25 mm diameter $(A_{st} = 1473 \text{ mm}^2)$ and two hanger bars of 12 mm diameter on the compression face.

7. Shear Reinforcements:

$$\tau_v = (V_u / b_w d) = [(123 \times 10^3)/(300 \times 500)] = 0.82 \text{ N/mm}^2$$

$$p_t = (100\ A_{st})/(b_w d) = [(100 \times 1473)/(300 \times 500)] = 0.98$$

From IS: 456 read out $\tau_c = 0.60\ N/mm^2$

Balance shear = $[V_{us} = V_u - (\tau_c b_w d)]$

$$= [123 - (0.60 \times 300 \times 500)\ 10 - 3]$$
$$= 33\ kN$$

Using 8 mm diameter two legged stirrups the spacing is,

$$s_v = \left[\frac{0.87 \times 415 \times 2 \times 50 \times 500}{33 \times 10^3}\right] = 547\ mm$$

But S_v = 0.75 d or 300 mm whichever is less,

$$= (0.75\ \times 500) = 375\ mm$$

Hence provide 8 mm diameter 2 legged stirrups at 300 mm centers throughout the length of the beam.

8. Check for Deflection Control:

$$p_t = (100A_{st})/(b_w d) = [(100 \times 1473)/(300 \times 500)] = 0.98$$
$$(b_w / b_f) = (300/2530) = 0.118$$

Refer the below figure and read out K_t = 2.00.

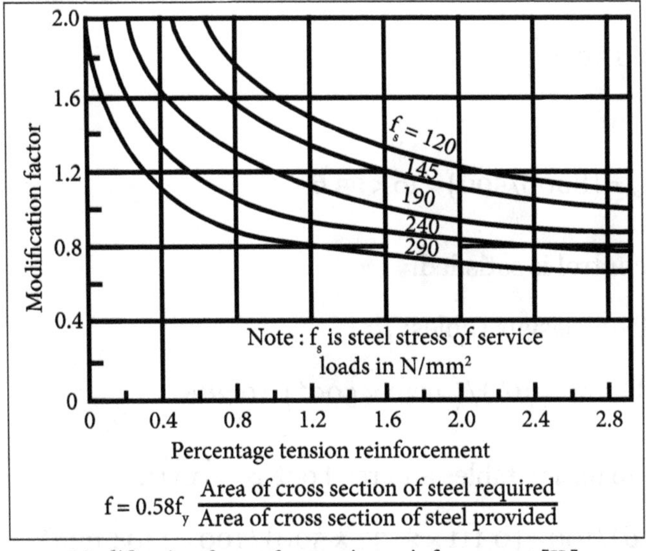

Modification factor for tension reinforcement [K_t].

Refer the below figure and read out $K_c = 1.00$.

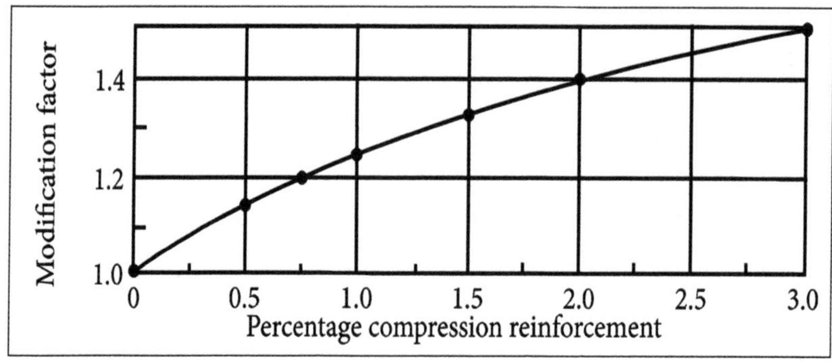

Modification factor for compression reinforcement [K_c].

Refer the below figure and read out $K_f = 0.80$.

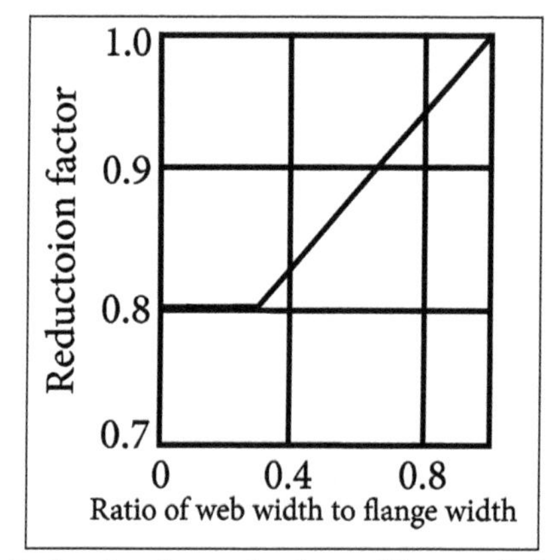

Reduction factor for flanged beams [K_f].

$$(L/d)_{max} = (L/d)_{basic} K_t \times K_c \times K_f$$

$$= [(16 \times 2 \times 1 \times 0.8)]$$

$$= 25.6$$

$$(L/d)_{provided} = (8000/500) = 16 < 25.6$$

Hence deflection control is satisfied.

9. Design using SP-16 Design Tables:

$$(M_u/b_f d^2) = 246 \times 10^6)/(2530 \times 500^2) = 0.388$$

Refer Table of SP-16 design tables and read out $p_t = 0.111$.

$$A_{st} = (p_t b_f d)/100 = (0\ 111 \times 2530 \times 500)/100 = 1404\ mm^2$$

The area of steel is the same as that computed using equations.

The reinforcement details in the tee beam is shown in the below figure:

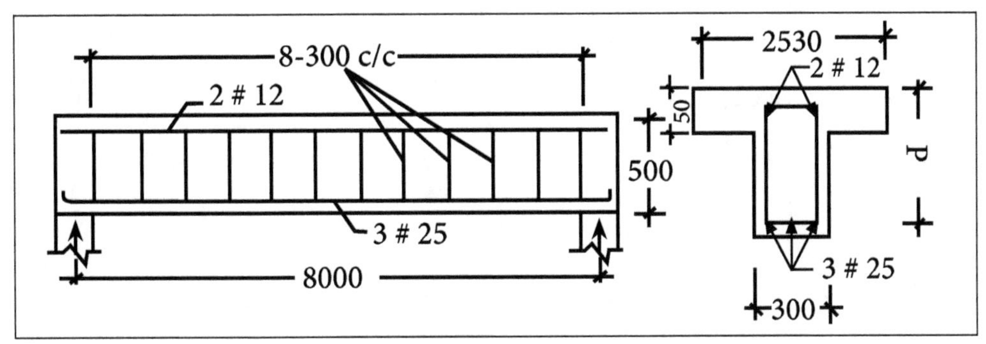

Reinforcement in tee beam.

2. Design of a L-beam for an office floor.

Solution:

Given:

Clear span = L = 8 m

Thickness of flange $= D_f = 150$ mm

Liver load $= g = 4$ kN/m^2

Spacing of beams = 3 m

f_{ck} 20 N/mm

f 415 N/mm

L-beams are monolithic with R.C. Columns.

Width of column = 300 mm

1. Cross Sectional Dimensions:

Since L-beam is subjected to bending, torsion and shear forces, assume a trial section having span/depth ratio of 12.

$$\therefore d = (8000/12) = 666 \text{ mm}$$

Adopt d = 700 mm

D = 750 mm

$b_w = 300$ mm

2. Effective Span:

Effective span is the least of the following values:

Centre to centre of supports = (8 + 0.3) = 8.3 m

Clear span + effective depth = (8 + 0.7) = 8.7 mm

$$\therefore\ L = 8.3 \text{ mm}$$

3. Loads:

Dead load of slab = (0.15 × 25 × 0.5 × 3) = 5.60 kN/m

Floor finish = (0.6 × 0.5 × 3) = 0.90

Self weight of rib = (0.3 × 0.6 × 25) = 4.50

Live load = (4 × 0.5 × 3) = 6.00

Total working load = w = 18.00 kN/m

4. Effective Flange Width:

Effective flange width (b_f) is the least of the following values:

$$b_f = (L_o/12) + b_w + 3\ D_f)$$

$$= (0.8300/12) + 300 + (3 \times 150) = 1442$$

$b_f - b_w + 0.5$ times the spacing between ribs,

$$b_f = 300 + (0.5 \times 2700) = 1650 \text{ mm}$$

$$\therefore\ b_f = 1442 \text{ mm}$$

5. Ultimate Bending Moment and Shear Forces:

At support section:

$$M_u = 1.5\ (17 \times 8.3^2)/12 = 147 \text{ kN.m}$$

$$V_u = 1.5(0.5 \times 17 \times 8.3) = 106 \text{ kN}$$

At centre of span section:

$$M_u = 1.5\ (17 \times 8.3^2)/24 = 73 \text{ kN.m}$$

6. Torsional Moments at Support Section:

Torsional moment is produced due to dead load of slab and live load on it.

(Working load/m - rib self weight) = (17 - 4.50) = 12.50 kN/m

∴ Total ultimate load on slab = 1.5 (12.50 × 8.3) = 156 kN

Total ultimate shear force = (0.5 × 156) = 78 kN

Distance of centroid of shear force from the center line of the beam.

= (0.5 × 1442- 150 = 571 mm

Ultimate torsional moment $= T_u = (78 \times 0.571) = 44.5$ kN.m

7. Equivalent Bending Moment and Shear Force:

According to IS: 456-2000, Clause 41.4.2, at the support section, the equivalent bending moment is compared as,

$$M_{el} = (M_u + M_t)$$

Where,

$$M_t = T_u = \left[\frac{1+(D/b)}{1.7}\right] = 44.5\left[\frac{1+(750/300)}{1.7}\right]$$
$$= 92 \text{ kN.m}$$

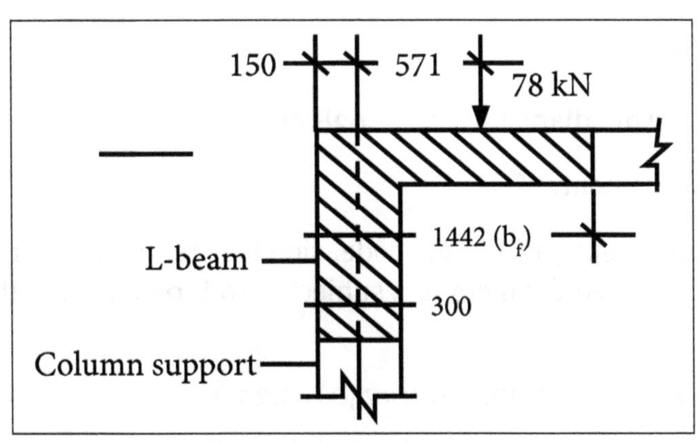

Eccentric load on L-beam.

$$\therefore M_{el} = (147 + 92) = 239 \text{ kN.m}$$
$$V_e = V_u + 1.6(T_u / b)$$
$$= 106 + 1.6(44.5/0.3)$$
$$= 344 \text{ kN}$$

8. Main Longitudinal Reinforcement:

Support section is designed as rectangular section to resist the hogging equivalent bending moment M_{el} = 239 kN.m.

$$M_{ulim} = 0-138\, f_{ck} b\, d^2$$
$$= \left(0.138 \times 20 \times 300 \times 700^2\right) 10^{-6}$$
$$= 405.7 \text{ kN.m}$$

Since $M_{el} < M_{ulim}$ the section is under reinforced:

$$M_{el} = 0.87\, f_y\, A_{st} d \left[1 - \left(\frac{A_{st}\, f_y}{b\, d\, f_{ck}}\right)\right]$$

$$\left(239 \times 10^6\right) = \left(0.87 \times 415\, A_{st} \times 700\right)\left[1 - \frac{415\, A_{st}}{\left(300 \times 700 \times 20\right)}\right]$$

Solving $A_{st} = 1056.3 \text{ mm}^2$

Provide 3 bars of 22 mm diameter on the tension side $\left(A_{st} = 1140 \text{ mm}^2\right)$.

Area of steel required at center of span to resist a moment of M_u = 73 kN.m will be less than the minimum given by:

$$A_{st(min)} = \left(\frac{0.85\, b_w d}{f_y}\right) = \left(\frac{0.85 \times 300 \times 700}{415}\right)$$
$$= 430 \text{ mm}^2$$

Provide 2 bars of 20 mm diameter $\left(A_{st} = 628 \text{ mm}^2\right)$.

9. Side Face Reinforcement:

According to Clause 26.5.1.7 of IS: 456 code, side face reinforcement of 0.1 per cent of web area is to be provided for members subjected to torsion, when the depth exceeds 450 mm.

$\therefore$ Area of reinforcement $= \left(0.001 \times 300 \times 750\right) = 225 \text{ mm}^2$

Provide 10 mm diameter bars (4 numbers) two on each face as horizontal reinforcement spaced 200 mm centers.

10. Shear Reinforcements:

$$\tau_{ve} = \left(\frac{V_e}{b_w d}\right) = \left(\frac{344 \times 10^3}{300 \times 700}\right) = 1.63 \text{ N/mm}^2$$

$$p_t = \left(\frac{100\ A_{st}}{b_w\ d}\right) = \left(\frac{100 \times 1140}{300 \times 700}\right) = 0.542$$

From Table (IS: 456) read out:

$$\tau_c = 0.49\ N/mm^2 < \tau_{ve}$$

Hence shear reinforcements are required.

Using 10 mm diameter two-legged stirrups with side covers of 25 mm and top and bottom covers of 50 mm, we have $b_1 = 250$ mm, $d_1 - 650$ mm, $A_{sv} = (2 \times 78.5) = 157\ mm^2$.

The spacing s_v is computed using the equations specified in Clause 41.4.3 of IS: 456-2000 code.

$$s_v = \left[\frac{(0.87\ f_y\ A_{sv}\ d_1)}{\left(\frac{T_u}{b_1}\right) + \left(\frac{V_u}{2.5}\right)}\right]$$

$$= \left[\frac{(0.87 \times 415 \times 157 \times 650)}{\left(\frac{44.5 \times 10^6}{250}\right) + \left(\frac{106 \times 10^3}{2.5}\right)}\right]$$

$$s_v = 167\ mm$$

Or

$$s_v = \left[\frac{A_{sv}\ 0.87\ f_y}{(\tau_{ve} - \tau_c)n}\right]$$

$$= \left[\frac{(157 \times 0.87 \times 415)}{(1.63 - 0.49)300}\right]$$

$$s_v = 165\ mm$$

Provide 10 mm diameter two-legged stirrups at a minimum spacing given by clause 26.5.1.7 of IS: 456. Adopt the minimum spacing based on shear and torsion considerations computed as $s_v = 160$ mm.

11. Check for Deflection Control:

$$p_t = \left(\frac{100 \times 1140}{300 \times 700}\right) = 0.54$$

$$p_c = \left(\frac{100 \times 628}{300 \times 700}\right) = 0.299$$

$$(b_w / b_f) = (300 / 1442) = 0.208$$

Refer the below Figure and read out K_t = 1.2.

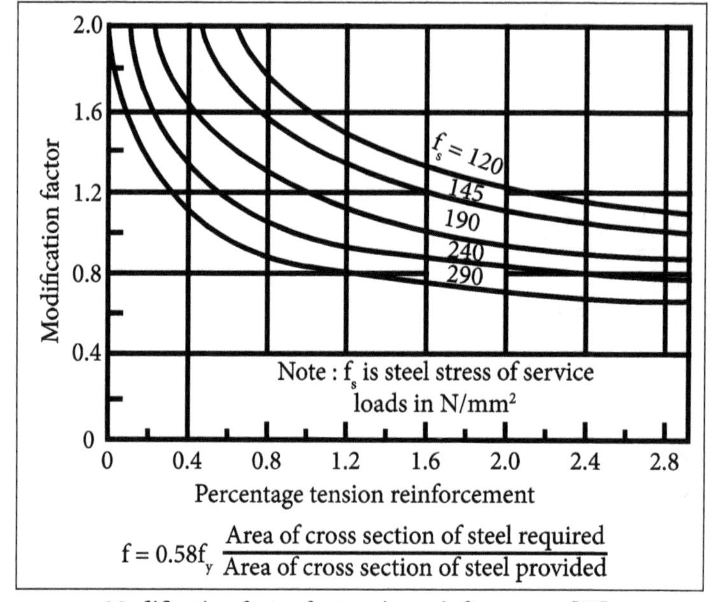

Modification factor for tension reinforcement [K_t].

Refer the below figure and read out K_c = 1.1.

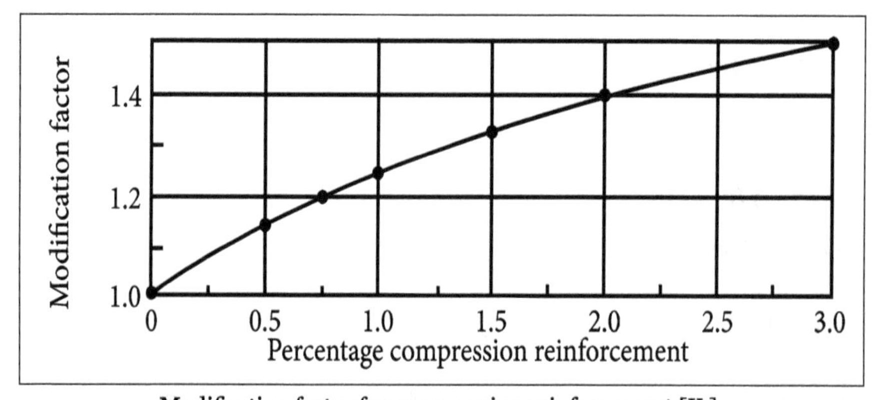

Modification factor for compression reinforcement [K_c].

Refer the below figure and read out K_f = 0.80.

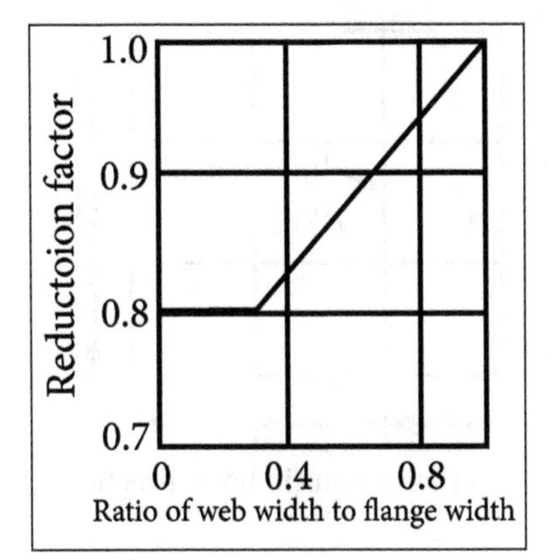

Reduction factor for flanged beams [K_f].

$$(L/d)_{max} = \left[(L/d)_{basic} \times K_t \times K_c \times K_f\right]$$

$$= \left[(9.20 \times 1.2 \times 1.1 \times 0.8)\right] = 21.12$$

$$(L/d)_{actual} = (8300/700) = 11.85 < 21.12$$

Hence deflection control is satisfied.

12. Reinforcement Details:

The below figure shows the details of reinforcements in the L-beam.

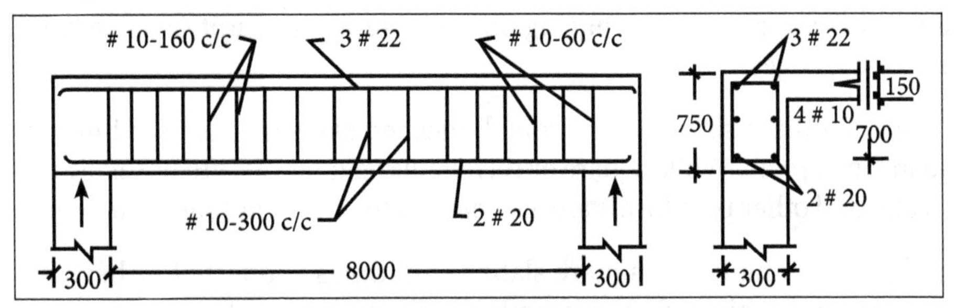

Reinforcement details in L-beam.

2.1.1 Design of Slab

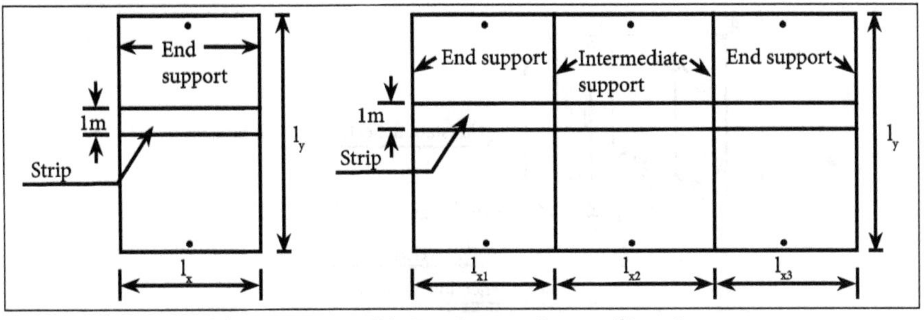

(a) Open span (b) Continuous in one direction.

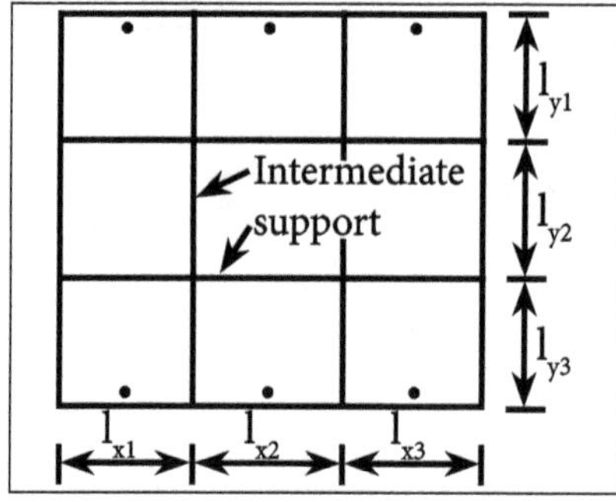

(c) Continuous in both direction.

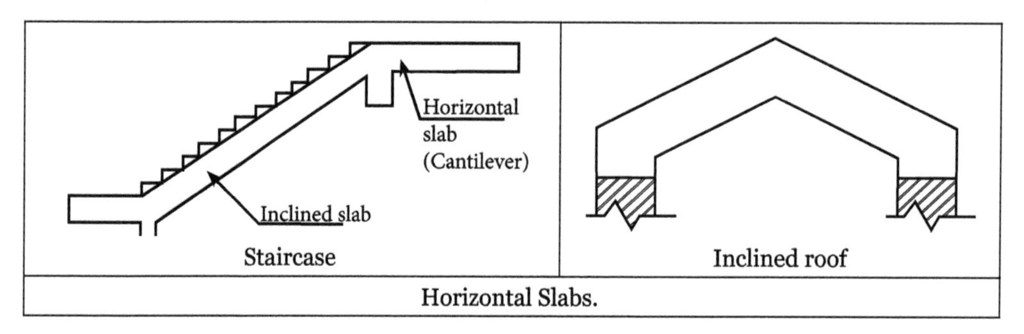

Horizontal Slabs.

Slabs, used in floors and roofs of buildings mostly integrated with the supporting beams, carry the distributed loads primarily by bending. A part of the integrated slab is considered as flange of T- or L-beams because of monolithic construction.

However, the remaining part of the slab needs design considerations. These slabs are either single span or continuous having different support conditions like fixed, hinged or free along the edges.

Though normally these slabs are horizontal, inclined slabs are also used in ramps, stair cases and inclined roofs. While square or rectangular plan forms are normally used, triangular, circular and other plan forms are also needed for different functional requirements.

Horizontal and rectangular or square slabs of buildings supported by beams in one or both directions and subjected to uniformly distributed vertical loadings.

One-way and Two-way Slabs

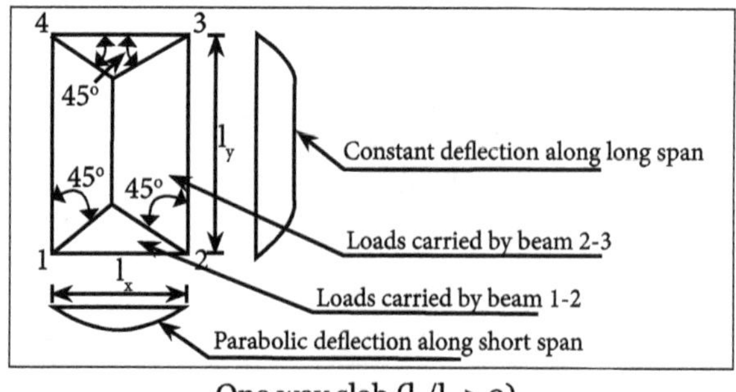

One way slab ($l_y/l_x > 2$).

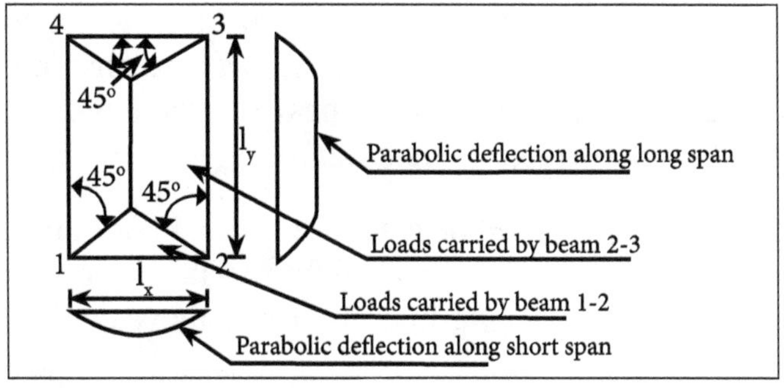

Two way slab ($l_y/l_x \leqq 2$).

The share of loads on beams in two perpendicular directions depends upon the aspect ratio l_y / l_x of the slab, l_x being the shorter span. For large values of l_y, the triangular area is much less than the trapezoidal area.

Hence, the share of loads on beams along shorter span will gradually reduce with increasing ratio of l_y / l_x. In such cases, it may be said that the loads are primarily taken by beams along longer span.

The deflection profiles of the slab along both directions are also shown in the figure. The deflection profile is found to be constant along the longer span except near the edges for the slab panel. These slabs are designated as one-way slabs as they span in one direction (shorter one) only for a large part of the slab when $l_y / l_x > 2$.

On the other hand, for square slabs of $l_y / l_x = 1$ and rectangular slabs of l_y / l_x up to 2, the deflection profiles in the two directions are parabolic. Thus, they are spanning in two directions and these slabs with l_y / l_x up to 2 are designated as two-way slabs, when supported on all edges.

It would be noted that an entirely one-way slab would need lack of support on short edges. Also, even for $l_y / l_x < 2$, absence of supports in two parallel edges will render the slab one-way. In figure b, the separating line at 45° is tentative serving purpose of design. Actually, this angle is a function of l_y / l_x.

Design Shear Strength of Concrete in Slabs

Experimental tests confirmed that the shear strength of solid slabs up to a depth of 300 mm is comparatively more than those of depth greater than 300 mm. Accordingly, Clause 40.2.1.1 of IS 456:2000 stipulates the values of a factor k to be multiplied with τ_c for different overall depths of slab.

Overall Depth of slab (mm)	300 or more	275	250	225	200	175	150 or less
k	1.00	1.05	1.10	1.15	1.20	1.25	1.30

Thin slabs, therefore, have more shear strength than that of thicker slabs. It is the normal practice to choose the depth of the slabs so that the concrete can resist the shear without any stirrups for slab subjected to uniformly distributed loads. However, for deck slabs, culverts, bridges and fly over, shear reinforcement should be provided as the loads are heavily concentrated in those slabs.

Though, the selection of depth should be made for normal floor and roof slabs to avoid stirrups, it is essential that the depth is checked for the shear for these slabs taking due consideration of enhanced shear strength as discussed above depending on the overall depth of the slabs.

Structural Analysis

One-way slabs subjected to mostly uniformly distributed vertical loads carry them primarily by bending in the shorter direction. Therefore, for the design, it is important to analyse the slab to find out the bending moment depending upon the supports. Moreover, the shear forces are also to be computed for such slabs.

These internal bending moments and shear forces can be determined using elastic method of analysis considering the slab as beam of unit width i.e. one metre. However, these values may also be determined with the help of the coefficients.

It is worth mentioning that these coefficients are applicable if the slab is of uniform cross-section and subjected to substantially uniformly distributed loads over three or more spans and the spans do not differ by more than fifteen per cent of the longer span.

It is also important to note that the average of the two values of the negative moment at the support should be considered for unequal spans or if the spans are not equally loaded.

Further, the redistribution of moments shall not be permitted to the values of moments obtained by employing the coefficients of bending moments.

For slabs built into a masonry wall developing only partial restraint, the negative moment at the face of the support should be taken as Wl/24, where W is the total design loads on unit width and l is the effective span.

The shear coefficients, in such a situation, may be increased by 0.05 at the end support.

Design Considerations

The primary design considerations of both one and two-way slabs are strength and deflection. The depth of the slab and areas of steel reinforcement are to be determined from these two aspects. However, the following aspects are to be decided first.

- Effective span (Clause 22.2 of IS 456:2000): The effective span of a slab depends on the boundary condition. Table gives the guidelines stipulated in Clause 22.2 of IS 456:2000 to determine the effective span of a slab.

S. number	Support condition	Effective span
1	Simply supported not built integrally with its supports.	Lesser of (i) clear span + effective depth of slab and (ii) centre to centre of supports.
2	Continuous when the width of the support is > lesser of 1/12th of clear span or 600 mm, (i) For end span with one end fixed and the other end continuous or for intermediate spans. (ii) For end span with one end free and the other end continuous. (iii) Spans with roller or rocker bearings.	(i) Clear span between the supports. (ii) Lesser of (a) clear span + half the effective depth of slab and (b) clear span + half the width of the discontinuous support. (iii) The distance between the centres of bearings.
3	Cantilever slab at the end of a continuous slab.	Length up to the centre of support.
4	Cantilever span	Length up to the face of the support + half the effective depth.
5	Frames	Centre to centre distance.

- Effective Span to Effective Depth Ratio (Clause.23.2.1(a-e) of IS 456:2000): The deflection of the slab can be kept under control if the ratios of effective span to effective depth of one-way slabs are taken up from the provisions in clause 23.2.1(a-e) of IS 456. These stipulations are for the beams and are also applicable for one-way slabs as they are designed considering them as beam of unit width.

- Nominal cover (clause 26.4 of IS 456): The nominal cover to be provided depends upon durability and fire resistance requirements. Appropriate value of the nominal cover is to be provided from these tables for the particular requirement of the structure.

- Minimum reinforcement (clause 26.5.2.1 of IS 456): Each for one and two-way slabs, the amount of minimum reinforcement in either direction shall not be less than 0.15 and 0.12 per cents of the total cross-sectional area for mild steel (Fe 250) and high strength deformed bars (Fe 415 and Fe 500)/ welded wire fabric, respectively.

- Maximum diameter of reinforcing bars (clause 26.5.2.2): The maximum diameter of reinforcing bars of one and two-way slabs shall not exceed one-eighth of the total depth of the slab.

- Maximum distance between bars (clause 26.3.3 of IS 456): The maximum horizontal distance between parallel main reinforcing bars shall be the lesser of (i) three times the effective depth or (ii) 300 mm. However, the same for secondary/distribution bars for temperature, shrinkage etc. shall be the lesser of (i) five times the effective depth or (ii) 450 mm.

Design of One Way Slabs

The reinforced concrete slabs supported on two opposite sides or on all four sides with the ratio of long to short span exceeding two are referred to as one way slabs. Slabs are designed as beams of unit width for a given type of loading and support conditions. Span/depth ratios specified in IS: 456- 2000 code for beam is also applicable for slabs.

The percentage of reinforcement in slabs is generally low in the range of 0.3 to 0.5 percent. Use of modification factor (K_f) for tension reinforcement results in the span/depth ratio in the range of 25 to 30 for one way slabs. The thickness of the slab should be such that the shear force at support is resisted by concrete alone without recourse to shear reinforcements.

Permissible shear stress is increased by the use of shear enhancement factor (k) specified in clause 40.2.1.1 of IS: 456-2000 code. Due to practical considerations, the depth of slab selected is generally greater than the minimum depth required for balanced section and hence the slab is under-reinforced.

Reinforcements in the slabs can be computed by using equations specified in Annexure G of the code or by using charts or tables of SP: 16. Slab designed for flexure is checked for shear stresses and limit state of deflection.

The procedure of the design of one-way slab is the same as that of beams. However, the amounts of reinforcing bars for one metre width of the slab is to be determined from either the governing design moments or from the requirement of minimum reinforcement. The different steps of the design are explained below.

Step 1: Selection of Preliminary Depth of Slab.

The depth of the slab shall be assumed from the span to effective depth ratios.

Step 2: Design Loads, Bending Moments and Shear Forces.

The total factored loads are to be determined adding the estimated dead load of the slab, load of the floor finish, given or assumed live loads, etc. after multiplying each of them with the respective partial safety factors.

Thereafter, the design positive and negative bending moments and shear forces are to be determined using the respective coefficients.

Step 3: Determination of the Effective and Total Depths of Slabs.

The effective depth of the slab shall be determined employing in equation and is given below as a ready reference here,

$$M_{u.lim} = R_{lim} bd^2$$

The value of b shall be taken as 1 metre.

The total depth of the slab shall then be determined adding appropriate nominal cover and half of the diameter of the larger bar if the bars are of different sizes.

Normally, the computed depth of the slab comes out to be much less than the assumed depth in Step 1. However, final selection of the depth shall be done after checking the depth for shear force.

Step 4: Depth of the Slab for Shear Force.

Theoretically, the depth of the slab can be checked for shear force if the design shear strength of concrete is known. Since this depends upon the percentage of tensile reinforcement, the design shear strength shall be assumed considering the lowest percentage of steel.

The value of τ_c shall be modified after knowing the multiplying factor k from the depth tentatively selected for the slab in Step 3. If necessary, the depth of the slab shall be modified.

Step 5: Determination of Areas of Steel.

Area of steel reinforcement along the direction of one-way slab should be determined employing equation,

$$M_u = 0.87 f_y A_{st} d \left\{1-(A_{st})(f_y)/(f_{ck})(b_d)\right\}$$

The above equation is applicable as the slab in most of the cases is under-reinforced due to the selection of depth larger than the computed value in Step 3. The area of steel so determined should be checked whether it is at least the minimum area of steel as mentioned in clause 26.5.2.1 of IS 456.

Alternatively, tables and charts of SP-16 may be used to determine the depth of the slab and the corresponding area of steel. Tables covering a wide range of grades of concrete and Chart 90 shall be used for determining the depth and reinforcement of slabs.

Tables of SP-16 take into consideration of maximum diameter of bars not exceeding one-eighth the depth of the slab. Zeros at the top right hand corner of these tables indicate the region where the percentage of reinforcement would exceed $p_{t,lim}$.

Similarly, zeros at the lower left hand corner indicate the region where the reinforcement is less than the minimum stipulated in the code. Therefore, no separate checking is needed for the allowable maximum diameter of the bars or the computed area of steel exceeding the minimum area of steel while using tables and charts of SP-16. The amount of steel reinforcement along the large span shall be the minimum amount of steel.

Step 6: Selection of Diameters and Spacing of Reinforcing Bars.

The diameter and spacing of bars are to be determined as per clause 26.5.2.2 and 26.3.3 of IS 456. As mentioned in Step 5, this step may be avoided when using the tables and charts of SP-16.

2.1.2 Detailing of Reinforcement

Figures shows the plan and section 1-1 of one-way continuous slab showing the different reinforcing bars in the discontinuous and continuous ends (DEP and CEP, respectively) of end panel and continuous end of adjacent panel (CAP).

The end panel has three bottom bars B1, B2 and B3 and four top bars T1, T2, T3 and T4. Only three bottom bars B4, B5 and B6 are shown in the adjacent panel. These bars are explained below for the three types of ends of the two panels.

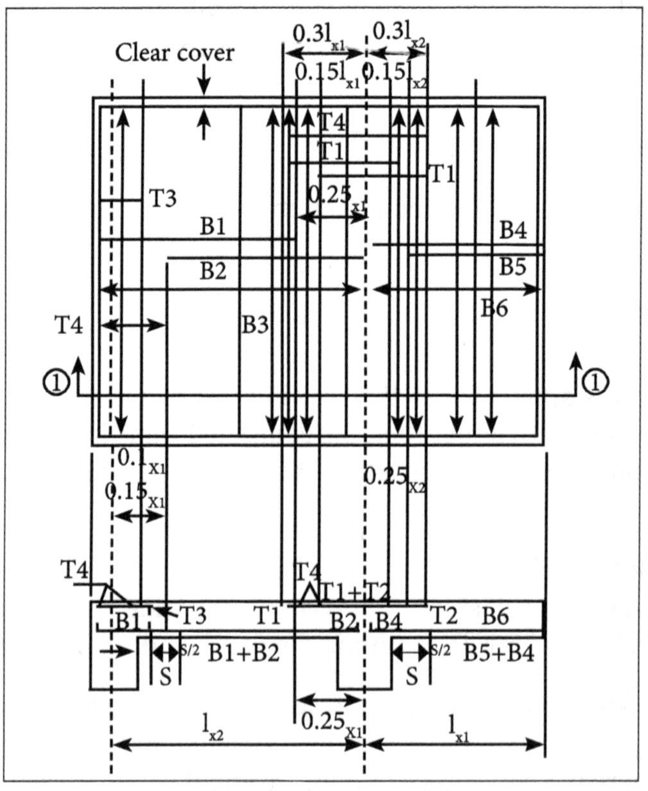

Reinforcement of one way slab.

S. No.	Bars	Panel	Along	Resisting moment
1.	B1,B2	DEP	X	+ 0.5 M_x for each,
2.	B3	DEP	y	Minimum steel
3.	B4,B5	CAP	X	+ 0.5 M_x for each,
4.	B6	CAP	y	Minimum steel
5.	T1,T2	CEP	X	+ 0.5 M_x for each,
6.	T3	DEP	X	+ 0.5 M_x
7.	T4	DEP	y	Minimum steel

- DEP = Discontinuous End Panel.
- CEP = Continuous End Panel.
- CAP = Continuous Adjacent Panel.

Discontinuous End Panel (DEP)

- Bottom steel bars B1 and B2 are alternately placed such that B1 bars are curtailed at a distance of 0.25 l_{x1} from the adjacent support and B2 bars are started from a distance of $0.15l_{x1}$ from the end support.
- Thus, both B1 and B2 bars are present in the middle zone covering $0.6l_{x1}$, each of which is designed to resist positive moment $0.5M_x$. These bars are along the direction of x and are present from one end to the other end of l_y.
- Bottom steel bars B3 are along the direction of y and cover the entire span l_{x1} having the minimum area of steel. The first bar shall be placed at a distance not exceeding s/2 from the left discontinuous support, where s is the spacing of these bars in y direction.
- Top bars T3 are along the direction of x for resisting the negative moment which is numerically equal to 50% of positive M_x. These bars are continuous up to a distance of $0.1l_{x1}$ from the centre of support at the discontinuous end.
- Top bars T4 are along the direction of y and provided up to a distance of $0.1l_{x1}$ from the centre of support at discontinuous end. These are to satisfy the requirement of minimum steel.

Continuous End Panel (CEP)

- Top bars T1 and T2 are along the direction of x and cover the entire l_y. They are designed for the maximum negative moment M_x and each has a capacity of $-0.5\ M_x$. Top bars T1 are continued up to a distance of $0.3l_{x1}$, while T2 bars are only up to a distance of $0.15l_{x1}$.

- Top bars T4 are along y and provided up to a distance of $0.3l_{x1}$ from the support. They are on the basis of minimum steel requirement.

Continuous Adjacent Panel (CAP)

- Bottom bars B4 and B5 are similar to B1 and B2 bars of (i) above.
- Bottom bars B6 are similar to B3 bars of (i) above.

Detailing is an art and hence structural requirement can be satisfied by more than one mode of detailing each valid and acceptable.

Let us design the one-way continuous slab subjected to uniformly distributed imposed loads of 5 kN/m² using M20 and Fe415. The load of floor finish is 1 kN/m². The span dimensions shown in the figure are effective spans. The width of beams at the support = 300 mm.

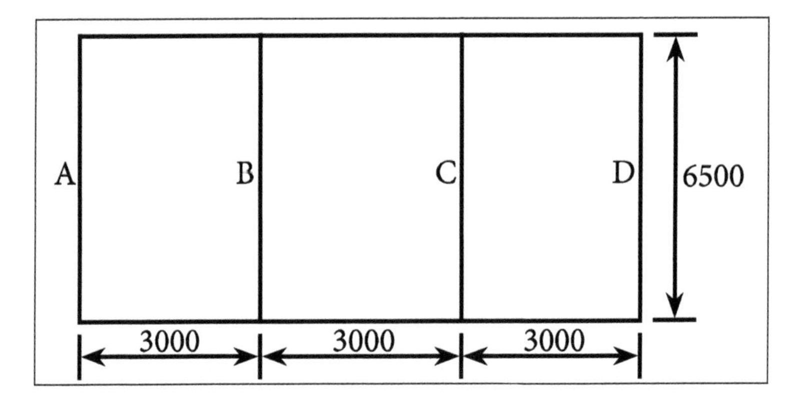

Solution:

Given:

Uniformly distributed imposed loads $= 5\ \text{kN}/\text{m}^2$

Load of finished floor $= 1\ \text{kN}/\text{m}^2$

Support = 300 mm

Step 1: Selection of Preliminary Depth of Slab.

The basic value of span to effective depth ratio for the slab having simple support at the end and continuous at the intermediate is (20+26)/2 = 23 (clause 23.2.1 of IS 456).

Modification factor with assumed p = 0.5 and $f_s = 240\ \text{N}/\text{mm}^2$ is obtained as 1.18 from figure of IS 456.

Therefore, the minimum effective depth = 3000/23(1.18) = 110.54 mm. Let us take the effective depth d = 115 mm and with 25 mm cover, the total depth D = 140 mm.

Step 2: Design Loads, Bending Moment and Shear Force.

Dead loads of slab of 1 m width = 0.14(25) = 3.5 kN/m

Dead load of floor finish =1.0 kN/m

Factored dead load = 1.5(4.5) = 6.75 kN/m

Factored live load = 1.5(5.0) = 7.50 kN/m

Total factored load = 14.25 kN/m

Maximum moments and shear are determined from the coefficients.

Maximum positive moment = 14.25(3)(3)/12 = 10.6875 kN

Maximum negative moment = 14.25(3)(3)/10 = 12.825 kN

Maximum shear $V_u = 14.25(3)(0.4) = 17.1$ kN

Step 3: Determination of Effective and Total Depths of Slab.

$$M_{u,lim} = R_{,lim} bd^2, \text{where } R_{,lim} \text{is } 2.76 \text{ N/mm}^2$$

So, $d = \{12.825(10^6)/(2.76)(1000)\}^{0.5} = 68.17$ mm

Since, the computed depth is much less than that determined in Step 1, keep D = 140 mm and d = 115 mm.

Step 4: Depth of Slab for Shear Force.

Table of IS 456 gives $\tau_c = 0.28 \text{ N/mm}^2$ for the lowest percentage of steel in the slab. Further for the total depth of 140 mm, let us use the coefficient k of clause 40.2.1.1 of IS 456 as 1.3 to get $\tau_c = k\,\tau_c == 1.3(0.28) = 0.364 \text{ N/mm}^2$.

Table of IS 456 gives $\tau_{c\,max} = 2.8 \text{ N/mm}^2$. For this problem $\tau_v = v_u / V_{bd} = 17.1/115$ $= 0.148 \text{ N/mm}^2$. Since, $\tau_v < \tau_c < \tau_{c.\,max}$, the effective depth d = 115 mm is acceptable.

Step 5: Determination of Areas of Steel.

$$M_u = 0.87\, f_y A_{st} d\, \{1 - (A_{st})(f_y)/(f_{ck})(bd)\}$$

For the maximum negative bending moment,

$$12825000 = 0.87(415)(A_{st})(115)\{1 - (A_{st})(415)/(1000)(115)(20)\}$$

$$\text{or } A_{st}^{2} = 5542.16\, A_{st} + 1711871.646 = 0$$

Solving the quadratic equation, we have the negative $A_{st} = 328.34\ mm^2$.

For the maximum positive bending moment,

$$10687500 = 0.87(415)\, A_{st}\, (115)\ \{1-(A_{st})(415)/(1000)(115)(20)\}$$

$$\text{Or } A_{st}^{2} = 5542.16\, A_{st} + 1426559.705 = 0$$

Solving the quadratic equation, we have the positive $A_{st} = 270.615\ mm^2$.

Alternative Approach:

Use of Table of SP-16:

For negative bending moment,

$$M_u / bd^2 = 0.9697$$

Table of SP-16 gives: $p_s = 0.2859$ (by linear interpolation). So, the area of negative steel = 0.2859(1000)(115)/100 = 328.785 mm^2.

For positive bending momenl,

$$M_u / bd^2 = 0.8081$$

Table of SP-16 gives: $p_s = 0.23543$ (by linear interpolation). So, the area of positive steel = 0.23543(1000)(115)/100 = 270.7445 mm^2.

Distribution Steel Bars along Longer Span l_y

Distribution steel area = Minimum steel area = $0.12(1000)(140)/100 = 168\ mm^2$. Since, both positive and negative areas of steel are higher than the minimum area, we provide:

- For negative steel: 10 mm diameter bars @ 230 mm c/c for which $A_{st} = 341\ mm^2$ giving $p_s = 0.2965$.
- For positive steel: 8 mm diameter bars @ 180 mm c/c for which $A_{st} = 279\ mm^2$ giving $p_s = 0.2426$.
- For distribution steel: Provide 8 mm diameter bars @ 250 mm c/c for which A_{st} (minimum) = 201 mm^2.

Step 6: Selection of Diameter and Spacing of Reinforcing Bars.

The diameter and spacing already selected in step 5 for main and distribution bars are checked below. For main bars (clause 26.3.3.b.1 of IS 456), the maximum spacing is the lesser of 3d and 300 mm i.e., 300 mm. For distribution bars (clause 26.3.3.b.2 of IS 456), the maximum spacing is the lesser of 5d or 450 mm i.e., 450 mm. Provided spacings, therefore, satisfy the requirements.

Maximum diameter of the bars (clause 26.5.2.2 of IS 456) shall not exceed 140/8 = 17 mm is also satisfied with the bar diameters selected here.

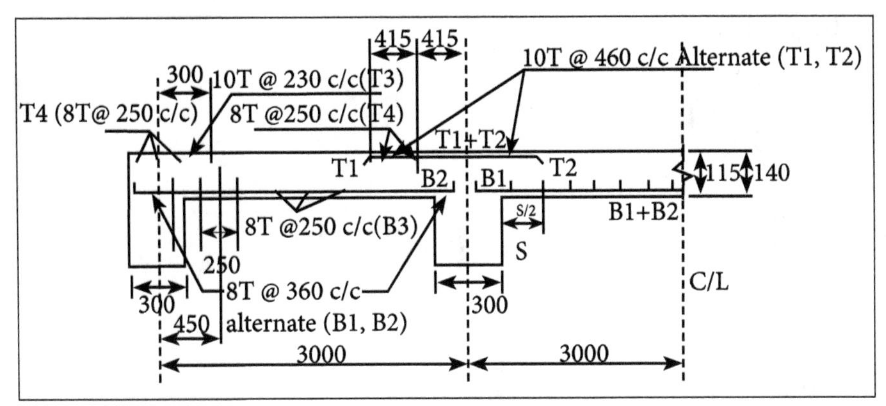

Reinforcement detail.

The figure presents the detailing of the reinforcement bars. The abbreviation B1 to B3 and T1 to T4 are the bottom and top bars, respectively which are shown for a typical one-way slab.

The above design and detailing assume absence of support along short edges. When supports along short edges exist and there is eventual clamping top reinforcement would be necessary at shorter supports also.

Problem

1. Let us design a one way slab having a clear span of about 2.5m and the slab supported on load bearing brick walls having 230mm thick. The residential floor has a load of 2 kN/m ^{2}. Use M_{20} grade concrete and Fe 415 HYSD bars.

Solution:

Given:

Clear span = 2.5 m.

Slab supported on load bearing brick walls 230 mm thick.

Loading: Residential floor, 2 kN / m^2.

Materials: M-20 grade concrete.

Fe-415 HYSD bars.

Formula to be used:

$$M = \left(0.125\ w\ L^2\right)$$

$$d = \sqrt{\frac{M}{Q\ b}}$$

$$V = (0.5\ w\ L)$$

$$A_{st} = \left(\frac{M}{\sigma_{st} \cdot j.d}\right)$$

$$s = \left(\frac{1000\ a_{st}}{A_{st}}\right)$$

$$\tau_v = \left(\frac{V}{b\ d}\right)$$

1. Allowable Stresses:

$$\sigma_{cbc} = 7\ N/mm^2$$

$$\sigma_{st} = 230\ N/mm^2$$

$$Q = 0.91$$

$$j = 0.90$$

2. Depth of Slab:

Assuming 0.4 per cent of reinforcement in the slab, the value of Kt using Fe-415 HYSD bars, is around 1.25.

Hence $(L/d) = (L/d)_{basic} \times K_{t} \times K_{c}$

$$= (20 \times 1.25 \times 1)$$
$$= 25$$

$$\therefore d = (2500\ /\ 25) = 100\ mm$$

Adopt d = 100 mm and overall depth =130 mm.

3. Effective Span:

Effective span is the least of:

Center to center of support = (2.5 + 0.23) = 2.73 m

Clear span + effective depth = (2.5 + 0.10) = 2.60 m

∴ Effective span = L = 2.60 m

4. Loads:

Self weight of slab = $(0.13 \times 25) = 3.25 \text{ kN/m}^2$

Live load on floor = 2.00 kN/m^2

Floor finishes = 0.75 kN/m^2

Total load = $w = 6.00 \text{ kN/m}^2$

Considering 1m width of slab, the uniformly distributed load is 6 kN/m^2 on an effective span of 2.60 m.

5. Bending Moments and Shear Forces:

$$M = (0.125\, w\, L^2) = (0.125 \times 6 \times 2.6^2) = 5.07 \text{ kN.m}$$

$$V = (0.5\, w\, L) = (0.5 \times 6 \times 2.6) = 7.80 \text{ kN}$$

6. Effective Depth:

$$d = \sqrt{\frac{M}{Q\, b}} = \sqrt{\frac{5.07 \times 10^6}{0.91 \times 10^3}} = 75 \text{ mm}$$

Effective depth adopted d = 100 mm, hence safe.

7. Main Reinforcements:

$$A_{st} = \left(\frac{M}{\sigma_{st} \cdot j.d}\right) = \left(\frac{5.07 \times 10}{230 \times 0.9 \times 100}\right) = 245 \text{ mm}$$

Minimum reinforcement $= (0.0012 \times 130 \times 1000) = 156 \text{ mm}^2 < 245 \text{ mm}^2$.

Spacing of 10 mm diameter bars is given by:

$$s = \left(\frac{1000\, a_{st}}{A_{st}}\right) = \left(\frac{1000 \times 79}{245}\right) = 322 \text{ mm}$$

Provide 10 mm diameter bars at 300 mm centers $\left(A_{st} = 262 \text{ mm}^2\right)$.

8. Distribution Reinforcement:

$$A_{st} = (0.0012 \times 1000 \times 130) = 156 \text{ mm}^2$$

Provide 8 mm diameter bars at 300 mm centers $\left(A_{st} = 167 \text{ mm}^2\right)$.

9. Check for Shear Stress:

$$\tau_v = \left(\frac{V}{b\,d}\right) = \left(\frac{7.80 \times 10^3}{10^3 \times 100}\right) = 0.078 \text{ N/mm}^2$$

Assuming 50 per cent of reinforcement to be bent up near supports, we have:

$$\left(\frac{100\,A_{st}}{b\,d}\right) = \left(\frac{100 \times 0.5 \times 262}{1000 \times 100}\right) = 0.131$$

From IS: 456-2000,

Interpolating permissible shear stress for solid slabs is:

$$(k \cdot \tau_c) = (1.30 \times 0.18) = 0.234 \text{ N/mm}^2 > \tau_v.$$

Hence shear stresses are within safe permissible limits.

10. Check for Deflection Control:

Percentage reinforcement $= p_t = \left(\frac{100 \times 262}{1000 \times 100}\right) = 0.262$

$$\text{For } p_t = 0.262,\ K_t = 1.6$$

$$(L/d)_{max} = (20 \times 1.6) = 32$$

$$(L/d)_{provided} = (2600/100) = 26 < 32, \text{ Hence safe.}$$

2. Let us design a one-way slab with a clear span of 3.5 m, supported on 200 mm thick concrete masonry walls to support a live load of 4 kN/m². Adopt M-20 grade concrete and Fe-415 HYSD bars.

Solution:

Given:

Clear span = 3.5 m

Width of support = 200 mm

Live load $= 4 \text{ kN} / \text{m}^2$

Floor finish $= 1 \text{ kN} / \text{m}^2$

$f_{ck} = 20 \text{ N} / \text{mm}^2$

$f_y = 415 \text{ N} / \text{mm}^2$

1. Depth of Slab:

Assume depth d = (span/25) = (3500/25) = 140 mm

Assuming a clear cover of 20 mm and using 10 mm diameter bars we have:

Effective depth = d = 140mm

Overall depth = D = 165mm

2. Effective Span:

The least value of:

Clear span + effective depth = 3.5 + 0.14 = 3.64 m

Centre to center of supports = 3.5 + 0.20 = 3.70 m

Hence L = 3.64 m

3. Loads:

Self weight of slab = (0.165 × 25) = 4.125 kN/m

Floor finish = 1.000

Live load = 4.000

Total service load = w = 9.125kN/m

Ultimate load $= w_u = (1.5 \times 9.125) = 13.69 \text{ kN} / \text{m}$

4. Ultimate Moments and Shear Forces:

$$M_u = (0.125\, w_u L^2) = (0.125 \times 13.69 \times 3.64^2) = 22.67 \text{ kN.m}$$

$$V_u = (0.5\, w_u L) = (0.5 \times 13.69 \times 3.64) = 24.92 \text{ kN}$$

5. Limiting Moment of Resistance:

$$M_{u,lim} = 0.138\, f_{ck}\, b\, d^2$$

$$= \left(0.138 \times 20 \times 10^3 \times 140^2\right)10^{-6}$$

$$= 54 \text{ kN.m}$$

Since $M_u < M_{u\,lim}$, section is under-reinforced.

6. Main Reinforcements:

$$M_u = 0.87\, f_y\, A_{st}\, d\left[1 - \left(\frac{A_{st}\, f_y}{b\, d\, f_{ck}}\right)\right]$$

$$\left(22.67 \times 10^6\right) = \left(0.87 \times 415\, A_{st} \times 140\right)\left[1 - \frac{415\, A_{st}}{\left(10^3 \times 140 \times 20\right)}\right]$$

Solving $A_{st} = 480 \text{ mm}^2$

Using 10 mm diameter bars, the spacing is:

$$s = \left(\frac{1000\, A_{st}}{0.4\, b}\right) = \left(\frac{100 \times 78.5}{480}\right) = 164 \text{ mm}$$

Adopt a spacing of 160 mm and alternate bars are bent up at supports.

7. Distribution Reinforcement:

$A_{st} = 0.12$ per cent of gross cross-sectional area

$$= \left(0.0012 \times 10^3 \times 165\right) = 198 \text{ mm}^2$$

Provide 8 mm diameter bars at 250 mm centres $\left(A_{st} = 201 \text{ mm}^2\right)$.

8. Check for Shear Stress:

$$\tau_v = \left(V_u / b\, d\right) = \left(24.92 \times 10^3\right) / \left(10^3 \times 130\right) = 0.178 \text{ N/mm}^2$$

$$p_t = \left(100\, A_{st} / bd\right) - \left(100 \times 0.5 \times 480\right) / \left(10^3 \times 140\right) = 0.17$$

Permissible shear stress in slab is,

$$k\tau_c = \left(1.27 \times 0.28\right) = 0.35 \text{ N/mm}^2 > \tau_v$$

Hence the shear stresses are within safe permissible limits.

9. Check for Deflection Control:

$$(L/d)_{max} = \left[(L/d)_{basic} \times K_t \times K_c \times K_f\right]$$

$$p_t = (100 \times 480)/(10^3 \times 140) = 0.34$$

Refer the below figure for $K_t = 1.40$, $K_c = 1.00$, $K_f = 1.00$

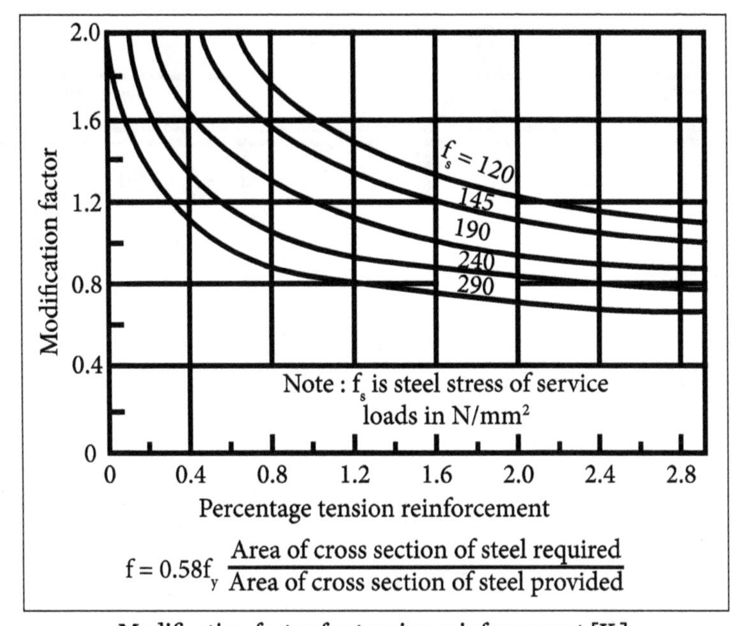

Modification factor for tension reinforcement [K_t].

$$\therefore (L/d)_{max} = \left[(20 \times 1.40 \times 1.00 \times 1.00)\right] = 28$$

$$(L/d)_{actual} = (3640/140) = 26 < 28$$

Hence the limit state of deflection is satisfied.

10. Design of Reinforcement Using SP-16 Design Tables:

$$\left(\frac{M_u}{b\,d^2}\right) = \left(\frac{22.67 \times 10^6}{10^3 \times 140^2}\right) = 115$$

Refer Table of SP-16 and read out the percentage of steel corresponding to $Fe-415\ N/mm^2$ and $f_{ck} = 20\ N/mm^2$.

$$P_t = 0.343$$

Hence $A_{st} = (p_t b\ d)/100$

$$= (0.343 \times 10^3 \times 140)/(100)$$
$$= 480\ mm^2.$$

11. Reinforcement Details:

The reinforcement details in the slab is shown in the below figure.

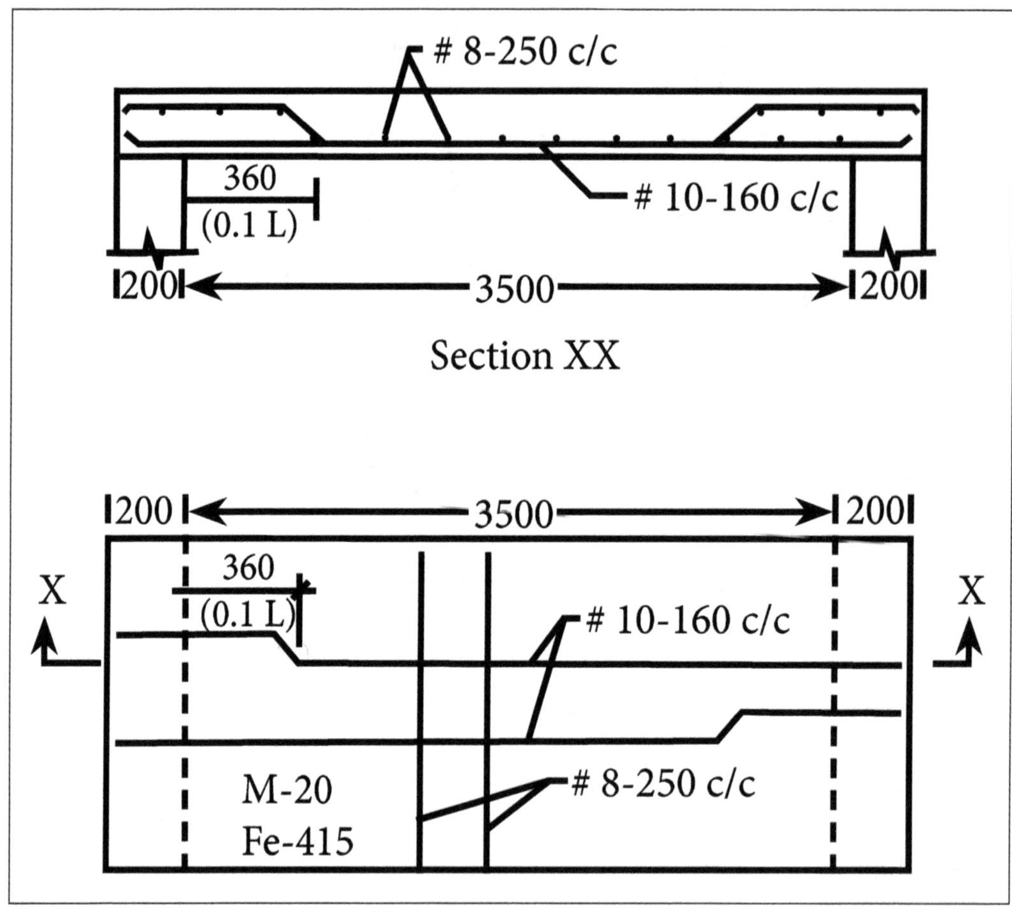

Details of reinforcements in one way slab.

2.1.3 Design of Two Way Slabs

Two-way slabs subjected mostly to uniformly distributed loads resist them primarily by bending about both the axes. However, as in the one-way slab, the depth of the two-way slabs should also be checked for the shear stresses to avoid any reinforcement for shear.

Moreover, these slabs should have sufficient depth for the control deflection. Thus, strength and deflection are the requirements of design of two-way slabs.

Design Shear Strength of Concrete

Design shear strength of concrete in two-way slabs is to be determined incorporating the multiplying factor k.

Structural Analysis

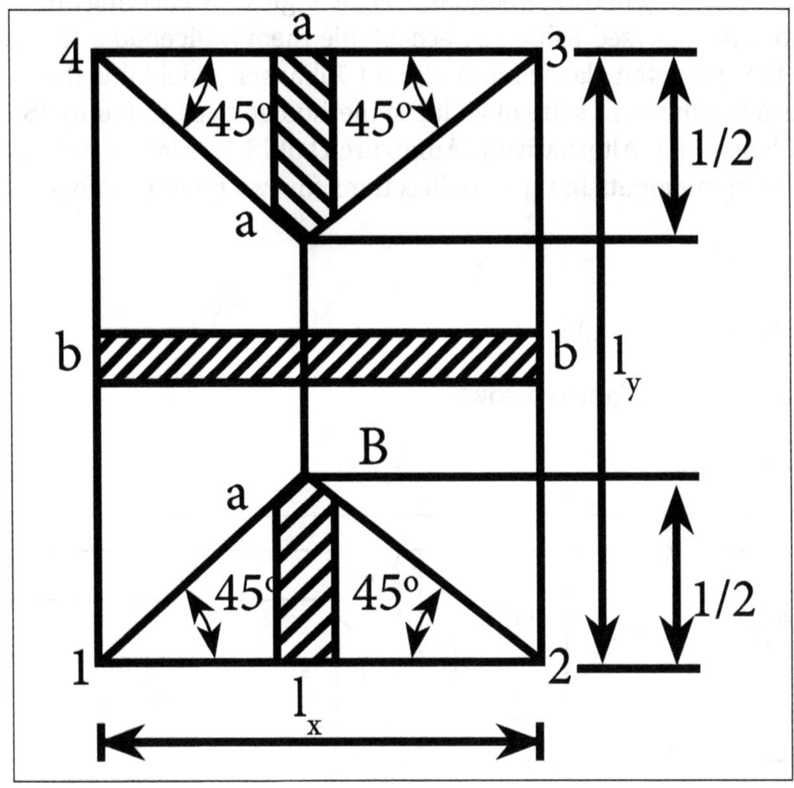

Strips for shear.

Computation of Shear Force

The two-way slab is divided into two trapezoidal and two triangular zones by drawing lines from each corner at an angle of 45°. The loads of triangular segment A will be transferred to beam 1-2 and the same of trapezoidal segment B will be beam 2-3.

The shear forces per unit width of the strips aa and bb are highest at the ends of strips. Moreover, the length of half the strip bb is equal to the length of the strip aa. Thus, the shear forces in both strips are equal and we can write,

$$V_u = W\,(l_x / 2)$$

Where,

W = Intensity of the uniformly distributed loads.

The nominal shear stress acting on the slab is then determined from,

$$\tau_v = V_u / bd$$

Computation of Bending Moments

Two-way slabs spanning in two directions at right angles and carrying uniformly distributed loads may be analysed using any acceptable theory. Pigeoud's or Wester-guard's theories are the suggested elastic methods and Johansen's yield line theory is the most commonly used in the limit state of collapse method and suggested by IS 456:2000 in the note of Clause 24.4. Alternatively, Annexure D of IS 456 can be employed to determine the bending moments in the two directions for two types of slabs:

- Restrained slabs.

- Simply supported slabs.

The two methods are explained below:

Restrained Slabs

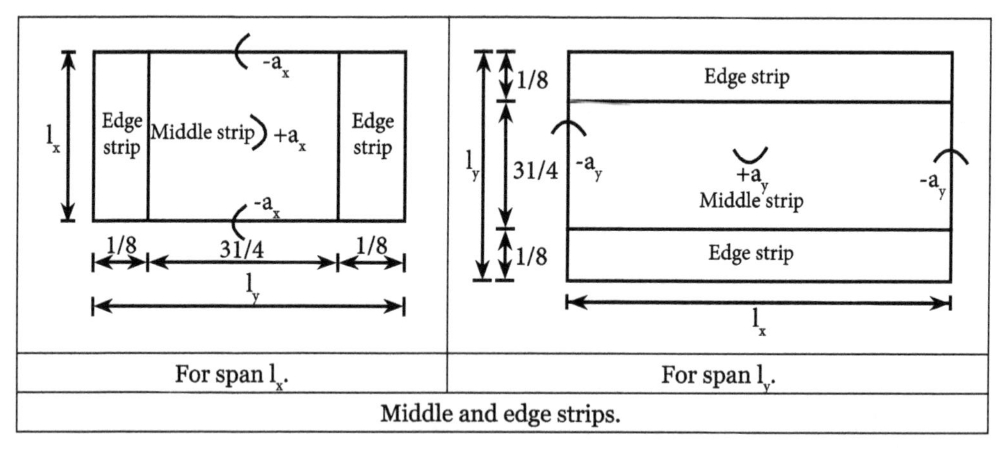

For span l_x. | For span l_y.

Middle and edge strips.

Restrained slabs are those whose corners are prevented from lifting due to effects of torsional moments. These torsional moments, however, are not computed as the amounts of reinforcement are determined from the computed areas of steel due to positive bending moments depending upon the intensity of torsional moments of different corners.

Thus, it is essential to determine the positive and negative bending moments in the two directions of restrained slabs depending on the various types of panels and the aspect ratio l_y / l_x. Restrained slabs are divided into two types of strips in each direction:

- One middle strip of width equal to three-quarters of the respective length of span in either directions.

- Two edge strips, each of width equal to 1/8th of the respective length of span in either directions. The maximum positive and negative moments per unit width in a slab are determined from,

$$M_x = \alpha_x . W_{lx}^{\ 2}$$
$$M_y = \alpha_x . W_{ly}^{\ 2}$$

Where, α_x, α_y are coefficients given in Table of IS 456:2000, Annex D, Clause D-1.1. Total design load per unit area is w and lengths of shorter and longer spans are represented by l_x and l_y, respectively.

The values of α_x and α_y, given in the table of IS 456, are for nine types of panels having eight aspect ratios of l_y / l_x from one to two at an interval of 0.1. The above maximum bending moments are applicable only to the middle strips and no redistribution shall be made.

Tension reinforcing bars for the positive and negative maximum moments are to be provided in the respective middle strips in each direction. Figure shows the positive and negative coefficients α_x and α_y.

The edge strips will have reinforcing bars parallel to that edge following the minimum amount as stipulated in IS 456.

Simply Supported Slabs

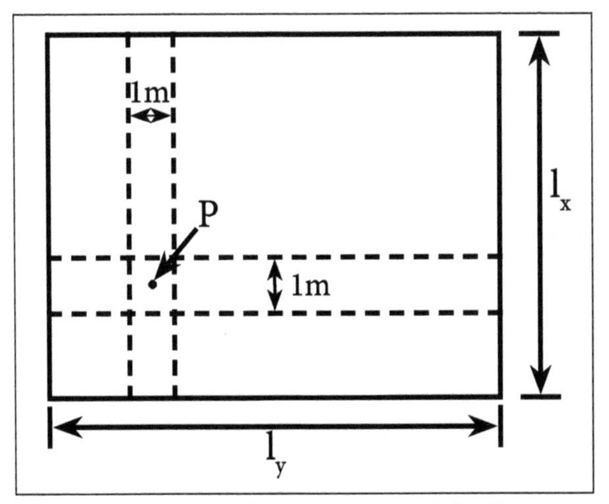

Interconnected two strips containing P.

The maximum moments per unit width of simply supported slabs, not having adequate provision to resist torsion at corners and to prevent the corners from lifting, are determined, where αx and αy are the respective coefficients of moments.

The coefficients αx and αy and of simply supported two-way slabs are derived from the

Grashoff-Rankine formula which is based on the consideration of the same deflection at any point P of two perpendicular interconnected strips containing the common point P of the two-way slab subjected to uniformly distributed loads.

Design Considerations

Effective Span to Effective Depth Ratio (Clause 24.1 of IS 456)

The following are the relevant provisions given in the notes 1 and 2 of clause 24.1 of IS 456:2000:

- The shorter of the two spans should be used to determine the span to effective depth ratio.
- For spans up to 3.5 m and with mild steel reinforcement, the span to overall depth ratios satisfying the limits of vertical deflection for loads up to $3 \text{ kN} / \text{m}^2$ are as follows:
  - Simply supported slabs-35.
  - Continuous slabs-40.
- The same ratios should be multiplied by 0.8 when high strength deformed bars (Fe 415) are used in the slabs.

Design Procedure of Two-Way Slabs

The procedure of the design of two-way slabs will have all the six steps as for the design of one-way slabs except that the bending moments and shear forces are determined by different methods for the two types of slab.

While the bending moments and shear forces are computed from the coefficients of IS 456 for the one-way slabs, the same are obtained for the bending moment in the two types of two-way slabs and the shear forces are computed.

Further, the restrained two-way slabs need adequate torsional reinforcing bars at the corners to prevent them from lifting. There are three types of corners having three different requirements.

Accordingly, the determination of torsional reinforcement is discussed in Step 7, as all the other six steps are common for the one and two-way slabs.

Step: Determination of Torsional Reinforcement.

Three types of corners, C1, C2 and C3, shown in figure, have three different requirements of torsion steel as mentioned below:

- At corner C1 where the slab is discontinuous on both sides, the torsion reinforcement shall consist of top and bottom bars each with layers of bar placed parallel to the sides of the slab and extending a minimum distance of one-fifth of the shorter span from the edges.

The amount of reinforcement in each of the four layers shall be 75 per cent of the area required for the maximum mid-span moment in the slab. This provision is given in clause D-1.8 of IS 456.

- At corner C2 contained by edges over one of which is continuous, the torsional reinforcement shall be half of the amount of (a) above. This provision is given in clause D-1.9 of IS 456.

- At corner C3 contained by edges over both of which the slab is continuous, torsional reinforcing bars need not be provided, as stipulated in clause D-1.10 of IS 456.

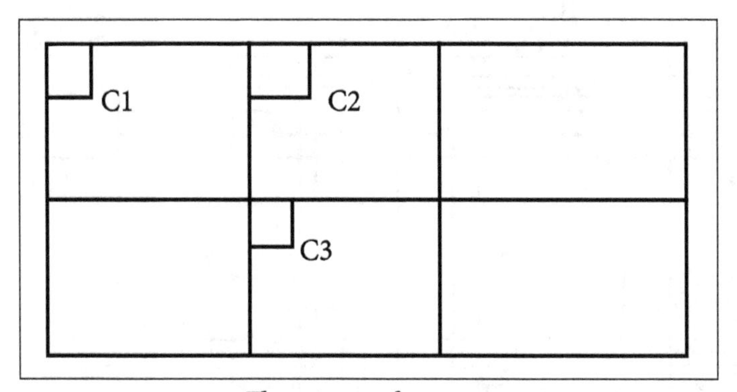

Three types of corners.

Detailing of Reinforcement

The above Step explains the method of determining the steel for corners of restrained slab depending on the type of corner. The detailing of torsional reinforcing bars is explained in above step. In the following detailing of reinforcing bars: Restrained slabs and Simply supported slabs.

These slabs are discussed separately for the bars either for the maximum positive or negative bending moments or to satisfy the requirement of minimum amount of steel.

Restrained Slabs

The maximum positive and negative moments per unit width of the slab calculated are applicable only to the respective middle strips. There shall be no redistribution of these moments. The reinforcing bars so calculated from the maximum moments are to be placed satisfying the following stipulations of IS 456.

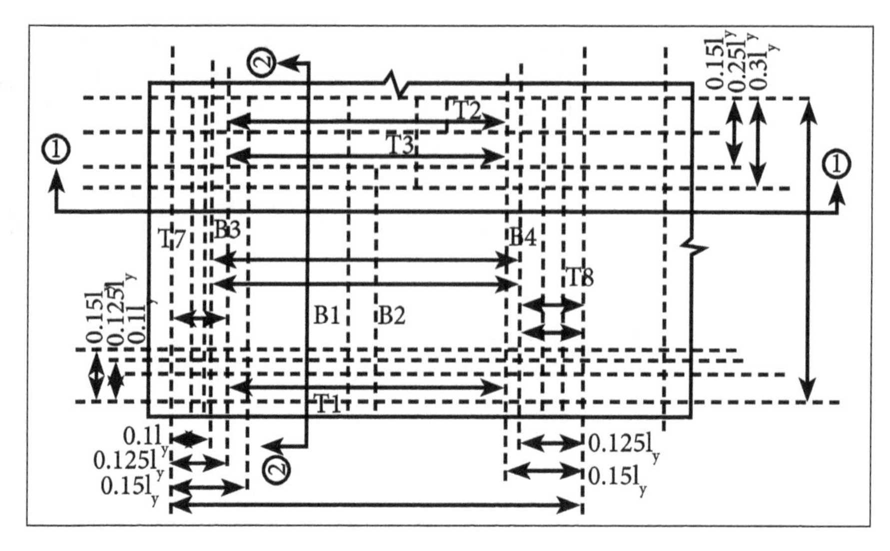

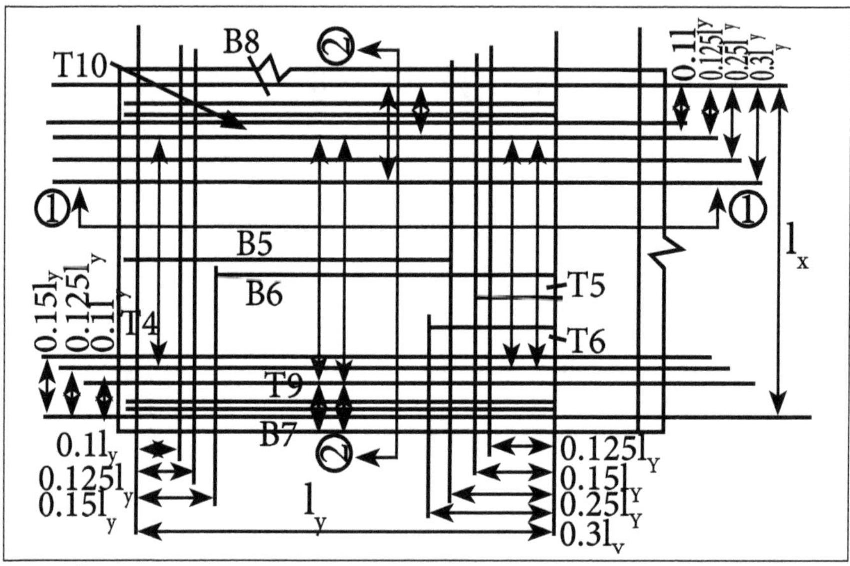

Bars along l_x only Reinforcement of two-way slab, $l_x < l_y$ (except torsion reinforcement).

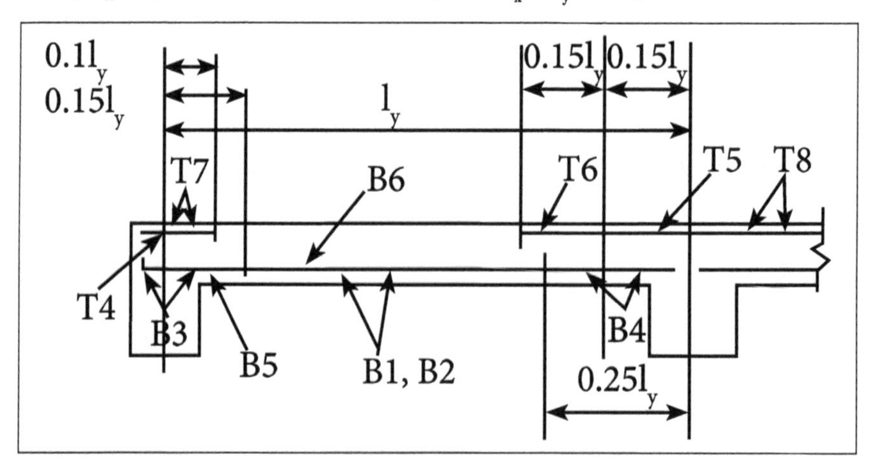

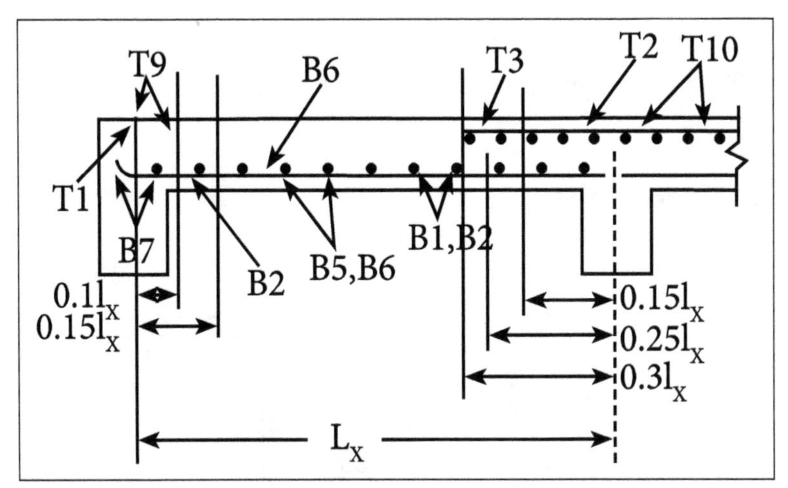

Reinforcement of Two-way slab, $l_x < l_y$ (except torsional reinforcement).

- Bottom tension reinforcement bars of mid-span in the middle strip shall extend in the lower part of the slab to within 0.25l of a continuous edge or 0.15l of a discontinuous edge (clause D-1.4 of IS 456). Bars marked as B1, B2, B5 and B6 are these bars.

- Top tension reinforcement bars over the continuous edges of middle strip shall extend in the upper part of the slab for a distance of 0.15lfrom the support and at least fifty per cent of these bars shall extend a distance of 0.3l (clause D-1.5 of IS 456). Bars marked as T2, T3, T5 and T6 are these bars.

- To resist the negative moment at a discontinuous edge depending on the degree of fixity at the edge of the slab, top tension reinforcement bars equal to 50% of that provided at mid-span shall extend 0.1l into the span (clause D-1.6 of IS 456). Bars marked as T1 and T4 in figure are these bars.

- The edge strip of each panel shall have reinforcing bars parallel to that edge satisfying the requirement of minimum amount and the requirements for torsion, explained in Step 7 (clause D-1.7 to D-1.10 of IS 456). The bottom and top bars of the edge strips are explained below.

- Bottom bars B3 and B4 are parallel to the edge along l_x for the edge strip for span l_y, satisfying the requirement of minimum amount of steel (clause D-1.7 of IS 456).

- Bottom bars B7 and B8 are parallel to the edge along l_y for the edge strip for span l_x, satisfying the requirement of minimum amount of steel (clause D-1.7 of IS 456).

- Top bars T7 and T8 are parallel to the edge along l_x for the edge strip for span l_y, satisfying the requirement of minimum amount of steel (clause D-1.7 of IS 456).

- Top bars T9 and T10 are parallel to the edge along l_y for the edge strip for span l_x, satisfying the requirement of minimum amount of steel (clause D-1.7 of IS 456).

Simply Supported Slabs

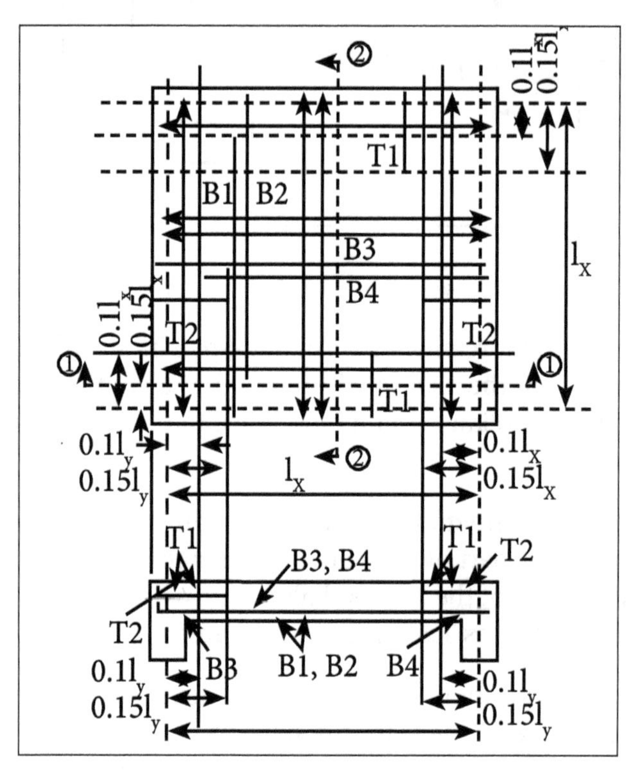

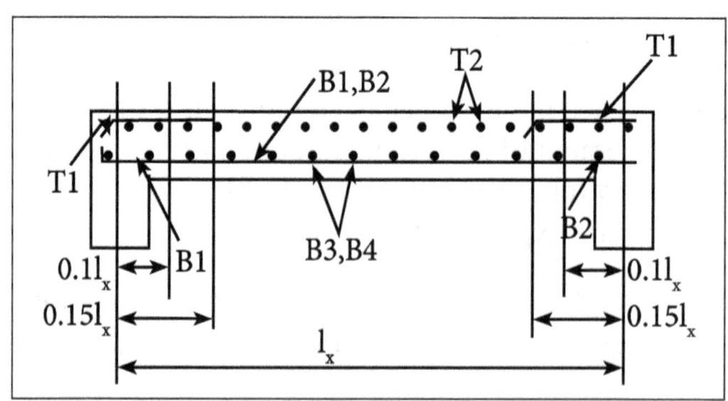

Simply supported two-way slab, corners not held down.

Figures present the detailing of reinforcing bars of simply supported slabs not having adequate provision to resist torsion at corners and to prevent corners from lifting.

Clause D-2.1 stipulates that fifty per cent of the tension reinforcement provided at mid-span should extend to the supports. The remaining fifty per cent should extend to within 0.1lx or 0.1ly of the support, as appropriate.

Let us design the slab panel 1 of the below figure subjected to factored live load of 8 kN/m² in addition to its dead load using M 20 and Fe 415. The load of floor finish is 1 kN/m². The spans shown in figure are effective spans. The corners of the slab are prevented from lifting.

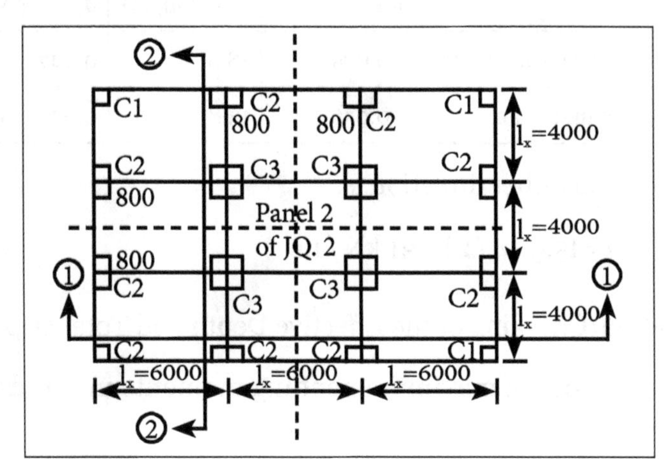

Solution:

Given:

Factored live load of 8 kN/m^2

Load of floor finish is 1 kN/m^2

Step 1: Selection of Preliminary Depth of Slab.

The span to depth ratio with Fe 415 is taken from clause 24.1, Note 2 of IS 456 as 0.8 (35 + 40) / 2 = 30. This gives the minimum effective depth d = 4000/30 = 133.33 mm, say 135 mm. The total depth D is thus 160 mm.

Step 2: Design Loads, Bending Moments and Shear Forces.

Dead load of slab (1 m width) $= 0.16(25) = 4.0 \text{ kN/m}^2$

Dead load of floor finish (given) $= 1.0 \text{ kN/m}^2$

Factored dead load $= 1.5(5) = 7.5 \text{ kN/m}^2$

Factored live load (given) $= 8.0 \text{ kN/m}^2$

Total factored load $= 15.5 \text{ kN/m}^2$

The coefficients of bending moments and the bending moments M_x and M_y per unit width (positive and negative) are determined as per clause D-1.1 and Table of IS 456 for the case 4,"Two adjacent edges discontinuous" and present in table. Then l_y / l_x, is 6/4 = 1.5.

Maximum Bending Moments

For	Short Span		Long Span	
	α_x	M_x(kNm/m)	α_y	M_y(kNm/m)
Negative moment at continuous edge	0.075	18.6	0.047	11.66
Positive moment at mid-span	0.056	13.89	0.035	8.68

Maximum shear force in either direction is,

$$V_u = w(l_x/2) = 15.5\ (4/2) = 31\ \text{kN/m}$$

Step 3: Determination/Checking of the Effective Depth and Total Depth of Slab.

Using the higher value of the maximum bending moments in x and y directions, we get,

$$M_{u,lim} = R_{lim} bd^2$$

$$\text{or } d = \left[(18.6)(10^6)/\{2.76(10^3)\}\right]^{1/2} = 82.09\ \text{mm},$$

Where, 2.76 N/mm² is the value of R_{lim}. Since, this effective depth is less than 135 mm assumed in Step 1, we retain d = 135 mm and D = 160 mm.

Step 4: Depth of Slab for Shear Force.

Table of IS 456 gives the value of $\tau_c = 0.28\ \text{N/mm}^2$ when the lowest percentage of steel is provided in the slab. However, this value needs to be modified by multiplying with k of clause 40.2.1.1 of IS 456. The value of k for the total depth of slab as 160 mm is 1.28. So, the value of τ_c is $1.28(0.28) = 0.3584\ \text{N/mm}^2$.

Table of IS 456 gives $\tau_c\ \text{max} = 2.8\ \text{N/mm}^2$. The computed shear stress $\tau_v = V_u/bd = 31/135 = 0.229\ \text{N/mm}^2$. Since, $\tau_v < \tau_c < \tau_c$ max, the effective depth of the slab as 135 mm and the total depth as 160 mm are safe.

Step 5: Determination of Areas of Steel.

The respective areas of steel in middle and edge strips are to be determined. Accordingly, the areas of steel for this problem are computed from the respective Tables of SP-16. Table of SP-16 is for the effective depth of 150 mm, while Table of SP-16 is for the effective depth of 175 mm.

The following results are, therefore, interpolated values obtained from the two tables of SP-16.

Reinforcing Bars

Particulars	Short Span lx			Long Span ly		
	Table No.	Mx(kN-m/m)	Diameter & spacing	Table No.	My(kNm/m)	Diameter & spacing
Negative moment at continuous edge	40,41	18.68 > 18.6	10 mm @ 200 mm c/c	40,41	12.314>11.66	8mm @ 200 mm c/c
Positive moment at mid-span	40,41	14.388 > 13.89	8mm @ 170 mm c/c	40,41	9.20 > 8.68	8mm @ 250 mm c/c

The minimum steel is determined from the stipulation of cl. 26.5.2.1 of IS 456 and is, $A_s = (0.12/100)(1000)(160) = 192\ mm^2$ and 8 mm bars @ 250 mm c/c $(= 201\ mm^2)$ is acceptable.

Step 6: Selection of Diameters and Spacing of Reinforcing Bars.

The advantages of using the tables of SP-16 are that the obtained values satisfy the requirements of diameters of bars and spacing. However, they are checked as ready reference here. Needless to mention that this step may be omitted in such a situation.

Maximum diameter allowed, as given in clause 26.5.2.2 of IS 456, is 160/8 = 20 mm, which is more that the diameters used here. The maximum spacing of main bars, as given in clause 26.3.3(1) of IS 456, is the lesser of 3(135) and 300 mm. This is also satisfied for all the bars. The maximum spacing of minimum steel (distribution bars) is the lesser of 5(135) and 450 mm. This is also satisfied.

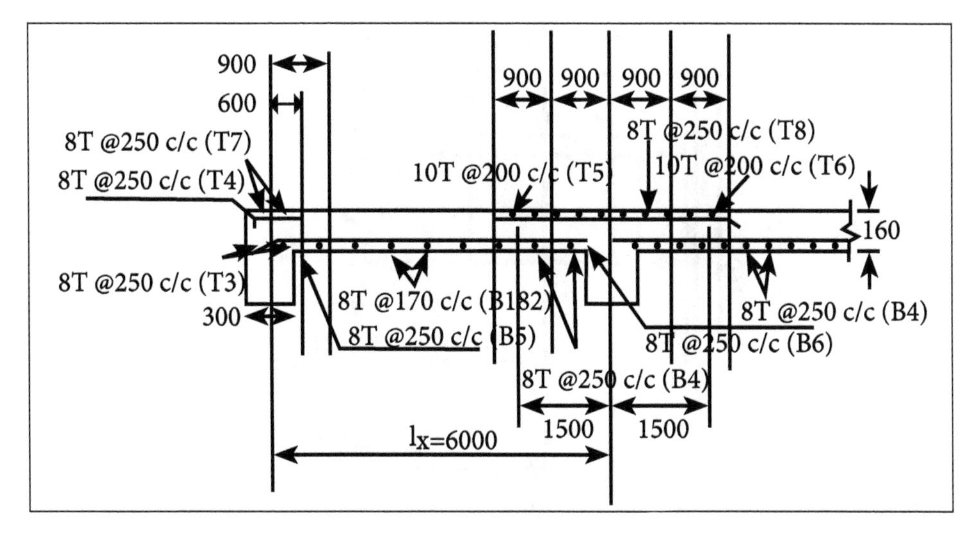

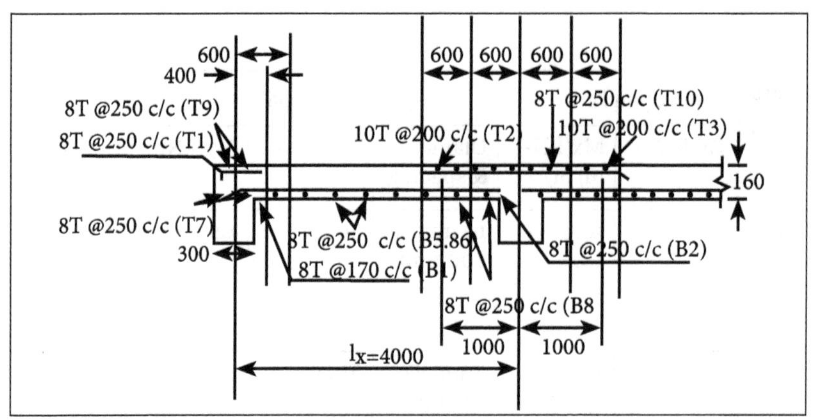

Reinforcement detail.

Step 7: Determination of Torsional Reinforcement.

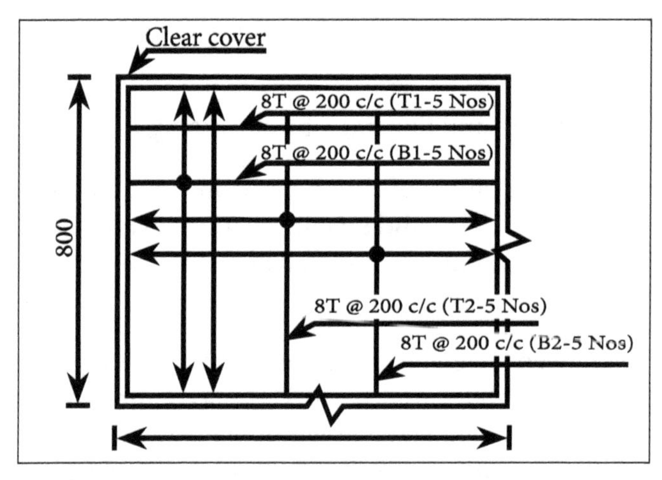

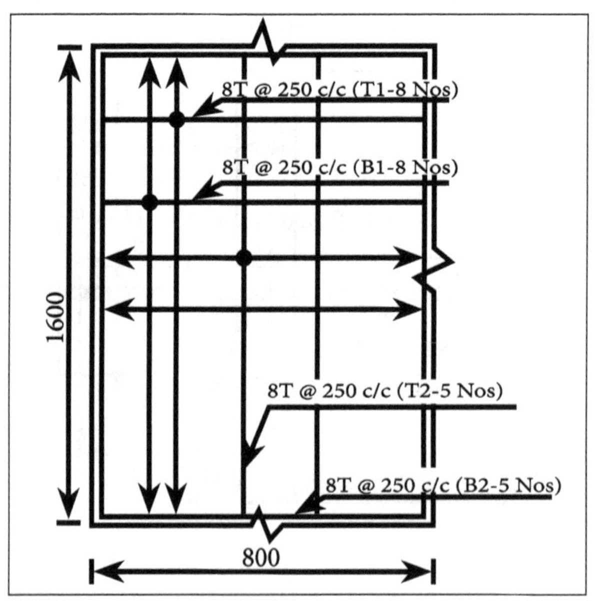

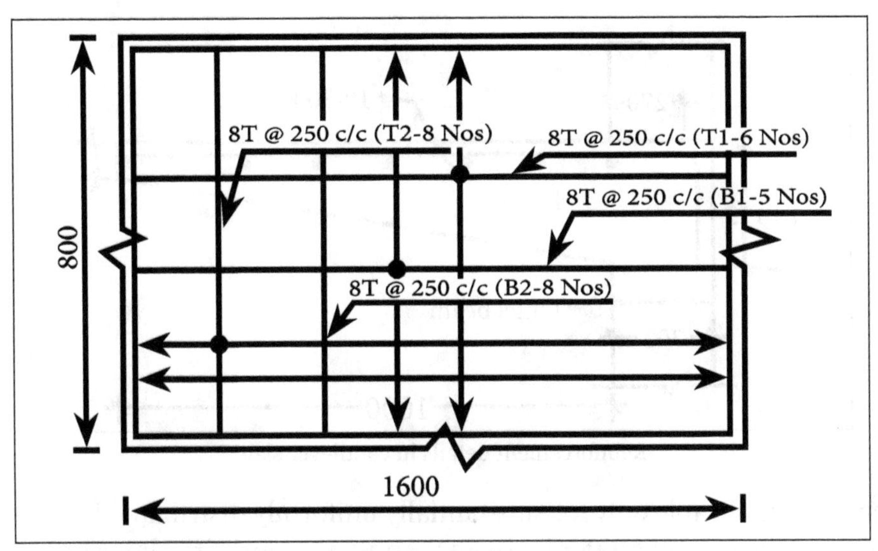

Corners C2: Torsion reinforcement bars.

Torsional reinforcing bars are determined for the three different types of corners. The length of torsional strip is 4000/5 = 800 mm and the bars are to be provided in four layers. Each layer will have 0.75 times the steel used for the maximum positive moment.

The C1 type of corners will have the full amount of torsional steel while C2 type of corners will have half of the amount provided in C1 type. The C3 type of corners do not need any torsional steel.

Type	Dimensions Along		Bar Diameter & Spacing	No. of Bars Along		Clause No of IS 456
	x(mm)	y(mm)		x	y	
C1	800	800	8mm @ 200 mm c/c	5	5	D-1.8
C2	800	1600	8mm @ 250 mm c/c	5	8	D-1.9
C3	1600	800	8mm @ 250 mm c/c	8	5	D-1.9

2.1.4 Design of Continuous Slabs

In multistorey buildings comprising tee beam and slab floors, the slabs are continuous over the beams which are spaced at regular intervals.

A simplified approach to design of continuous slabs is presented in IS: 456-2000 code.

Here moment and shear force coefficients are recommended for computation of moments and shear forces which are the same as that for continuous beams.

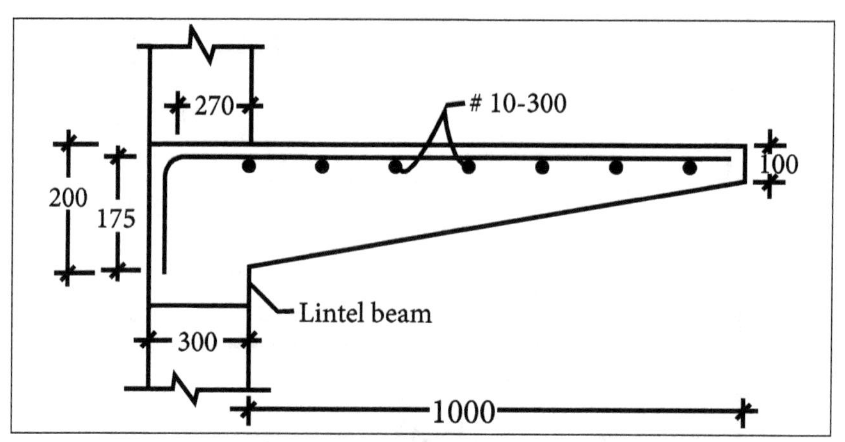

Reinforcement details in cantilever slab.

Coefficients are applicable only for substantially uniformly distributed loads over three or more spans which do not differ by more than 15 per cent of the longest span.

When coefficients specified in the tables of the code are used for computation of bending moments, redistribution of moments is not permitted. If the spans are significantly different, a rigorous analysis using the traditional methods such as moment distribution is made to compute the design maximum moments and shear forces.

Let us design a one way slab for an office floor which is continuous over tee beams spaced at 3.5 m intervals. Assume a live load of 4 kN/m² and adopt M_{20} grade concrete and Fe-415 HYSD bars.

Solution:

Given:

$$L = 3.5 \text{ m}$$

$$q = 4 \text{ kN/m}^2$$

$$f_{ck} = 20 \text{ N/mm}^2$$

$$f_y = 415 \text{ N/mm}^2$$

1. Depth of Slab:

Assuming a span/depth ratio of 26 (Clause 23.2.1 of IS: 456): Effective depth = d = (span/26) = (3500/26) = 135 mm.

Adopt d - 140 mm

D = 165 mm

2. Loads:

Self weight of slab $= (0.165 \times 25) = 4.125\ \text{kN/m}^2$

Finishes $= 0.875\ \text{kN/m}^2$

Total dead load (g) $= 5.000\ \text{kN/m}^2$

Service live load (q) $= 4\ \text{kN/m}^2$

3. Bending Moments and Shear Forces:

Referring to the tables of IS: 456-2000 code, maximum negative B.M. at support next to the end support is:

$$M_u(-ve) = 1.5\left[\frac{g\,L^2}{10} + \frac{q\,L^2}{9}\right]$$
$$= 1.5\left[\frac{5 \times 3.5^2}{10} + \frac{4 \times 3.5^2}{9}\right]$$
$$= 17.35\ \text{kN.m}$$

Positive B.M. at centre of span,

$$M_u(+ve) = 1.5\left[\frac{g\,L^2}{12} + \frac{q\,L^2}{10}\right]$$
$$= 1.5\left[\frac{5 \times 3.5^2}{12} + \frac{4 \times 3.5^2}{10}\right]$$
$$= 15\ \text{kN.m}$$

Maximum shear force at the support section is:

$$V_u = 1.5 \times 0.6\,(g+q)\,L$$
$$= (1.5 \times 0.6)\,(5+4)\,3.5$$
$$= 28.35\ \text{kN}$$

4. Check the Depth of Slab:

$$= \left(0.138 \times 20 \times 10^3 \times 140^2\right) 10^{-6}$$
$$= 54.1\ \text{kN.m}$$

Since $M_u < M_{u\,lim}$, section is under-reinforced.

5. Reinforcements:

$$M_u = 0.87\, f_y\, A_{st}\, d \left[1 - \left(\frac{A_{st}\, f_y}{b\, d\, f_{ck}}\right)\right]$$

$$\left(17.35 \times 10^6\right) = \left(0.87 \times 415\, A_{st} \times 140\right)\left[1 - \frac{415\, A_{st}}{\left(10^3 \times 140 \times 20\right)}\right]$$

Solving $A_{st} = 360\ mm^2/m$.

Provide 10 mm diameter bars at 150 mm centers $\left(A_{st} = 524\ mm^2\right)$. The same reinforcement is provided for positive B.M. at mid span.

Distribution steel = $\left(0.0012 \times 10^3 \times 165\right) = 198\ mm^2$

Provide 10 mm diameter bars at 300 mm centers $\left(A_{st} = 262\ mm^2\right)$.

6. Check for Shear Stress:

$$\tau_v = \left(v_u / b\, d\right)\ \left(28.35 \times 10^3\right)/\left(10^3 \times 140\right) = 0.20 N/mm^2$$

$$p_t = \left(100 A_{st}\right)/\left(bd\right) = \left(100 \times 262\right)/\left(10^3 \times 140\right) = 0.187$$

Refer Table of IS: 456 and read out:

$$k\, \tau_c = \left(1.27 \times 0.30\right) = 0.38\ N/mm^2$$

Since $\tau_v < \tau_c$, the slab is safe against shear stresses.

7. Check for Deflection Control:

Considering the end and interior spans,

$$\left(L/d\right)_{max} = \left[\left(L/d\right)_{basic} \times K_t \times K_c \times K_f\right] \text{ Also } K_c = K_f = 1.00$$

$$p_t = \left(\frac{100 \times 393}{10^3 \times 140}\right) = 0.28$$

From the figure, read out $K_t = 1.5$

$$\therefore \left(L/d\right)_{max} = \left(\frac{20 + 26}{2}\right) 1.5 = 34.5$$

$$\left(L/d\right)_{actual} = \left(3500/140\right) = 25 < 34.5$$

Hence the slab is safe against deflection control.

8. Reinforcement Details:

The reinforcement details in the continuous slab is shown in the below figure.

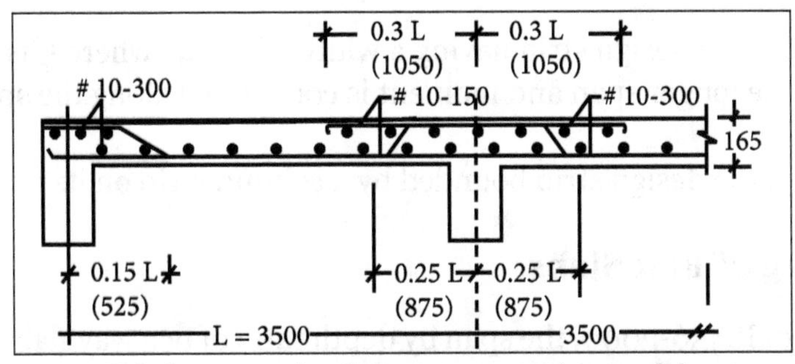

Reinforcement details in one way continuous slab.

Design of Flat Slabs

A flat slab is a typical type of construction in which a reinforced slab is built monolithically with the supporting columns and is reinforced in two or more directions without any provision of beams.

Components of Flat Slab

- Drop of flat slab.
- Capital or column head & Panel.

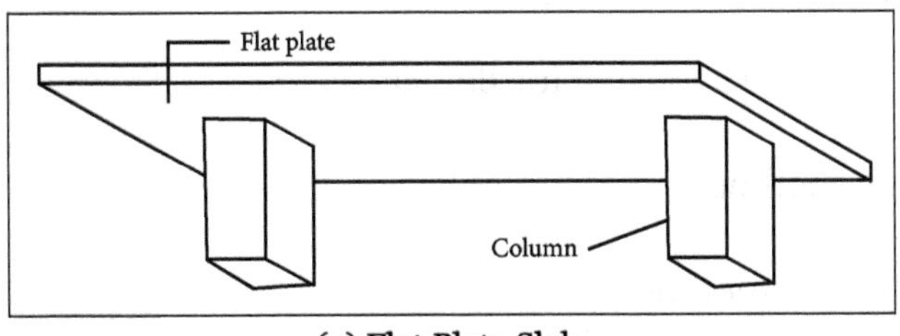

(a) Flat Plate Slab.

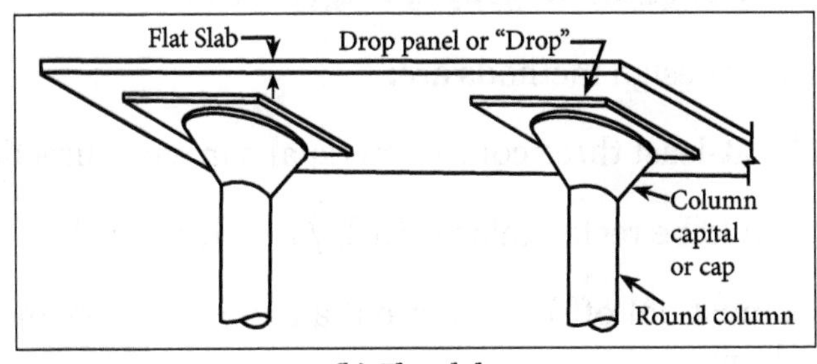

(b) Flat slab.

Types of Flat Slabs

- Slabs without drops and column heads.

- Slabs without drops.
- Slab with drops and column with column head.

Column Strip: It is the design strip having a width of $l_2/4$, where l_2 is the span transverse to l_1. l_2 is the longer span and moment is considered along the span l_1.

Middle Strip: It is the design strip bounded by a column strip on its opposite sides.

Proportioning of Flat Slabs

As per clause 31 of IS456-2000, the span by depth ratio of two way slab is applicable for flat slabs and the values can be (l/d)modified by 0.9 for flat slabs with drops.

Take l/d as 32 for HYSD bars.

As per ACI – The drop thickness should not be less than 100mm or (Thickness of slab)/4.

While calculating span by depth ratio, longer span is used. The thickness of slab should not be less than 125mm.

The purpose of column drop is to reduce the shear stress and also reduces the reinforcement in the column strip. The increase in column diameter at the head flaring of column head takes care of punching shear developed at a distance of d/2 all around the junction between the slab and column head.

Two methods of design are available for flat slabs:

- Direct design method.
- Equivalent frame method.

Direct Design Method: (Clause 31.4.1, IS456-2000)

Requirements for direct design methods are:

- There must be at-least three continuous spans in each direction.
- The panels should be rectangular with $l_y/l_x = l_2/l_1$ ratio < 2.
- The columns must not offset by more than 10% of the span from either of the successive columns.
- Successive span length in each direction must not differ by more than one third of longer span.
- Design live load must not exceed 3 times the designed dead load.

Design Procedure

As per Clause 31.4.2.2 IS456-2000, the total moment for a span bounded by columns laterally is given by,

$$M_o = Wl_o / 2$$

where,

M_o - The sum of positive and negative moment in each direction.

W - The total design load covered on an area L_2L_1.

$$W = w \times L_2 \times L_n$$

This moment is distributed for the column strip and middle strip.

Moments	Column strip	Middle strip
Negative moment (65%)	65 × 0.75 = 49%	65 × 0.2 = 15%
Positive moment (35%)	35 × 0.6 = 21%	35 × 0.4= 15%

Moment Distribution for Interior Panel

$$M_o = \frac{WL_o}{8}$$

Also L_o should not be less than 0.65 times of L_1 $(L_n > 0.65L_1)$.

Assumptions made in equivalent frame method:

- The structure is considered to be made of equivalent frames longitudinally and transversely.
- Each frame is analyzed by any established method like moment distribution method.
- The relative stiffness is computed by assuming gross cross section of the concrete alone in the calculation of the moment of inertia.
- Any variation of moment of inertia along the axis of the slab on account of provision of drops should be considered.

Flat Slab [Exterior Panel]

Stiffness of slab and column $= \frac{4\,EI}{L}$

Where,

$$I = bd^3/12 \text{ (or) } \pi d^4/64$$

$$E = 5000\sqrt{f_{ck}},$$

α_c is checked with α_c minimum given in Table of IS456. From Clause 31.4.3.3, the interior and exterior negative moments and the positive moments are found.

Interior negative design moment is expressed as,

$$0.75 - \frac{0.10}{1 + \frac{1}{\alpha_c}}$$

Where,

$$\alpha_c = \frac{\sum K_c}{K_z}$$

Interior positive design moment is,

$$0.63 - \frac{0.28}{1 + \frac{1}{\alpha_c}}$$

Exterior negative design moment is,

$$\frac{0.65}{1 + \frac{1}{\alpha_c}}$$

The distribution of interior negative moment for column strip and middle strip is in the ratio 3:1 (0.75 : 0.25). The exterior negative moment is fully taken by the column strip. The distribution of positive moment in column strip and middle strip is in the ratio 1.5 : 1 (0.6 : 0.4).

1. Let us design a flat slab system (interior panel) to suit the following data:

Size of the floor = 20 × 30m

Column interval = 5m c/c

Live load on slab = 5kN/m²

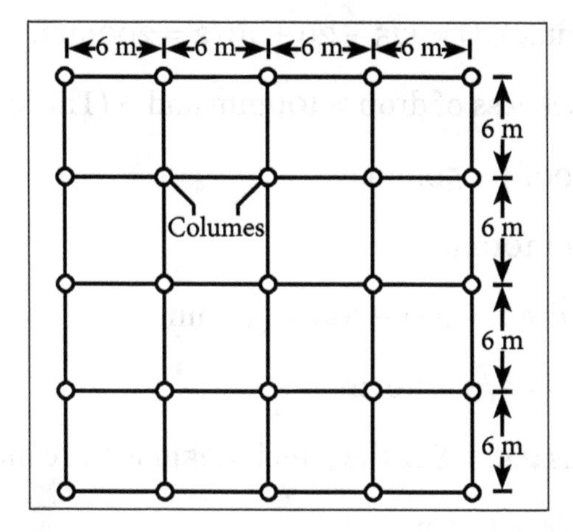

Materials used are Fe415 HYSD bars and M20 concrete.

Solution:

Given:

Size of the floor = 20 × 30m

Column interval = 5m c/c

Live load on slab = $5kN/m^2$

Materials used are Fe415 HYSD bars and M20 concrete.

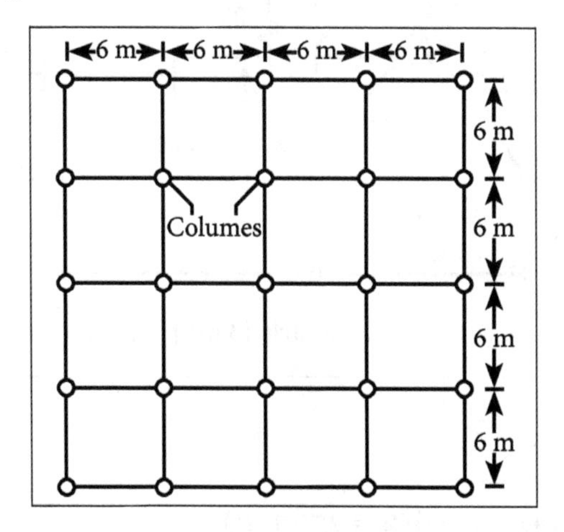

Proportioning of Flat Slab

Assume l/d as 32,

$d = 5000/32$

$d = 156.25mm$

$d = 175mm$ (assume), $D = 175 + 20 + 10/2 = 200mm$.

As per ACI code, the thickness of drop > 100mm and > (Thickness of slab)/4.

Therefore, 100mm or 200/4 = 50mm.

Provide a column drop of 100mm.

Overall depth of slab at drop = 200 + 100 = 300mm

Length of the drop > L/3 = 5/3 = 1.67m.

Provide length of drop as 2.5m. For the panel, 1.25m is the contribution of drop.

Column head = L/4 = 5/4 = 1.25m,

$$L_1 = L_2 = 5m$$

$$L_n = L_2 - D = 5 - 1.25 = 3.75m$$

As per code,

$$M_o = \frac{WL_o}{8}$$

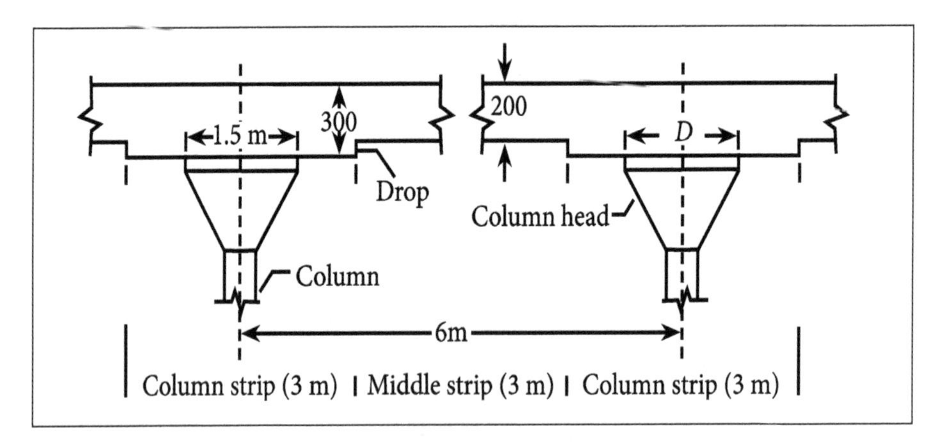

Loading on Slab

(Average thickness = (300 + 200)/2 = 250mm)

Self weight of slab $= 25 \times 0.25 = 6.25kN/m^2$

Live load $= 5\ kN/m^2$

Floor finish $= 0.75\ kN/m^2$

Total $= 12\ kN/m^2$

Factored load $= 1.5 \times 12 = 18 \text{ kN/m}^2$

$$W = w_u \times L_2 \times L_n = 18 \times 5 \times 3.75 = 337.5\text{kN}$$

Total moment on slab panel = (337.5 × 3.75)/8 = 158.203kNm.

Distribution of Moment

Moments	Column Strip	Middle Strip
Negative moment (65%)	65 × 0.75 = 49% 0.49 × 158.2 = 77.52kNm	65 × 0.2 = 15% 0.15 × 158.2 = 23.73kNm
Positive moment (35%)	35 × 0.6 = 21% 0.21 x 158.2 = 33.22kNm	35 × 0.4 = 15% 0.15 × 158.2 = 23.73kNm

2.1.5 Check for Depth Adopted

Column Strip

$$M_u = 0.138.f_{ck}.b.d^2b = 2.5m$$

$$77.5 \times 10^6 = 0.138 \times 20 \times 2.5 \times 1000 \times d^2$$

$$d = 105.98\text{mm} \approx 106\text{mm} < 275\text{mm}.$$

Middle Strip

$$M_u = 0.138.f_{ck}.b.d^2b = 2.5m$$

$$23.73 \times 10^6 = 0.138 \times 20 \times 2.5 \times 1000 \times d^2$$

$$d = 58.68\text{mm} \approx 59\text{mm} < 175\text{mm}.$$

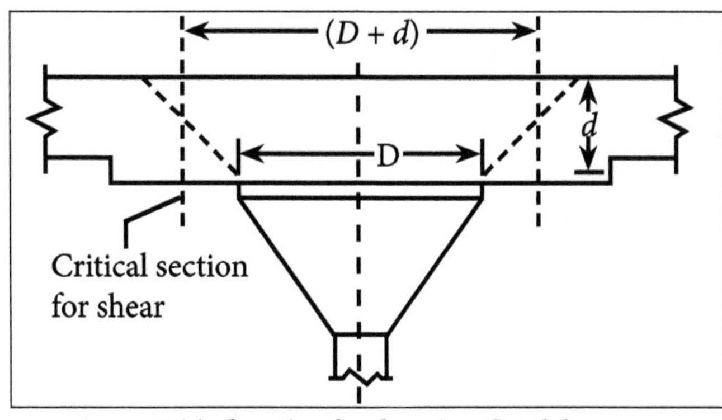

Critical section for shear in a flat slab.

Check for Punching Shear

The slab is checked for punching shear at a distance of d/2 all around the face of the column head. The load on the slab panel excluding the circular area of diameter (D + d) is the punching shear force.

Shear force = Total Load – (Load on circular area)

$$= (18 \times 5 \times 5) - \left(\pi(D+d)^2 / 4\right) \times wn$$

$$= 417.12 \text{ kN}$$

Shear force along the perimeter of the circular area $= \dfrac{\text{Shear Force}}{\pi(D+d)}$

$$= 87.06 \text{ kN}$$

Nominal shear stress: (b = 1m)

$$\varsigma_v = \frac{V_u}{b.D} = \frac{87.06 \times 10^3}{1000 \times 275} = 0.317 \text{ N/mm}^2$$

Design shear stress: $\zeta_c = K.\zeta_c'$

Where, K = (0.5 + β) ≤1 = 1.5 ≤ 1

$$\zeta_c' = 0.25.\, f_{ck} = 1.118 \text{ N/mm}^2$$

$$\zeta = 1 \times 1.118 = 1.118 \text{ N/mm}$$

$$\zeta_v < \zeta_c$$

Safe in shear.

Reinforcement

Column strip: (b = 2.5m), (d = 275mm)

Negative moment $= 77.5 \times 10^6$ Nmm

$$M_u = 0.87.f_y.A_{st}.\left(d - 0.42\left(\frac{0.87.f_y\, A_{st}}{0.36.f_{ck}.b}\right)\right)$$

Or

$$K = \frac{M_u}{bd^2}$$

Take p_t from SP 16,

$$77.6 \times 10^6 = 99.29 \times 103.A_{st} - 3.04.Ast^2$$

$$A_{st} = 800.16 \text{ mm}^2$$

Required 10mm @ 240mm c/c.

Minimum A_{st} : 0.12% of c/s = $0.12/100 \times 1000 \times 275 = 825 \text{ mm}^2$

Provide 10mm @ 230mm c/c.

Positive moment = 33.2 kNm

$$A_{st} = 337.876 \text{ mm}^2$$

Provide 8mm @ 370mm c/c.

Minimum steel: Provide 10mm @ 230mm c/c.

Middle strip: (b = 2.5m), (d = 175mm)

Negative and positive moment: 23.7 kNm

$$A_{st} = 382.6 \text{ mm}^2$$

$$A_{st\ min}. = (0.12/100 \times 1000 \times 2500 \times 175) = 525 \text{ mm}^2$$

Provide 8mm @ 230mm c/c.

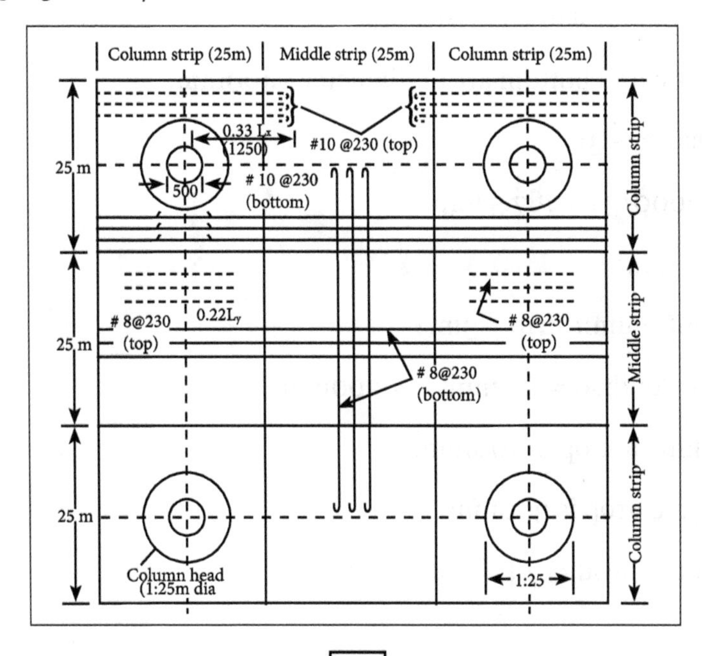

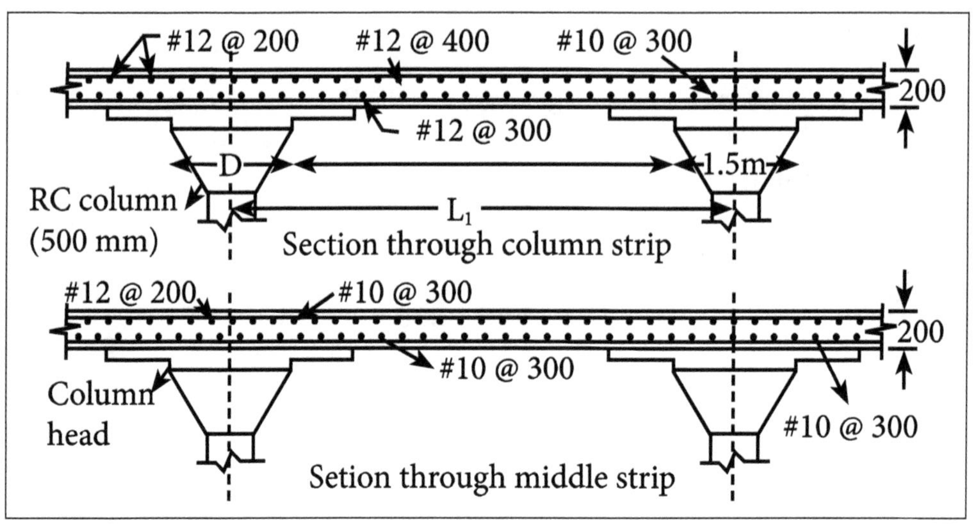

Reinforcement detail.

2. Let us design an exterior panel of a flat slab floor system of size 24m × 24m, divided into panels 6m x 6m size. The live load on the slab is 5 kN/m² and the columns at top and bottom are at diameter 400mm. Height of each storey is 3m. Use M_{20} concrete and Fe_{415} steel.

Solution:

Given:

Flat slab floor system

Size - 24m × 24m

Live load on the slab - 5 kN/m²

The columns at top and bottom are at diameter - 400mm.

Height of each storey - 3m.

$l/d = 32 \rightarrow d = 6000/32 = 187.5$ mm

Length of drop ≥ 3m.

Length of drop = Column strip = 3m.

Assume effective depth, d = 175mm, D = 200mm.

As per ACI, Assume a drop of 100mm.

Depth of slab at the drop is 300mm.

Diameter of column head = $l/4 = 6/4 = 1.5$m.

Loading on Slab

Self weight of slab = $(0.2+0.3)/2\times25=0.25\times25=6.25\ \text{kN}/\text{m}^2$

Live load = $5\ \text{kN}/\text{m}^2$

Floor finish = $0.75\ \text{kN}/\text{m}^2$

Total = $12\ \text{kN}/\text{m}^2$

Factored load $=1.5\times12=18\ \text{kN}/\text{m}^2$

To find the value of $\alpha c=\dfrac{\sum K_c}{K_s}$ as per Cl.3.4.6 of IS456-2000,

αc = flexural stiffness of column and slab.

ΣK_c = summation of flexural stiffness of columns above and below.

ΣK_s = summation of flexural stiffness of slab.

$$\sum K_c=2\left(\frac{4EI}{L}\right)=2\left(\frac{4xExI_c}{L_c}\right)=\frac{2x4x22.360x10^3x1.25x10^9}{3000}$$

Where,

$I=\pi d^4/64=\pi\times400^4/64$

$E=5000\sqrt{f_{ck}}=22.3606\times10^3$

$$\Sigma K_s=\frac{4xEx6000x(250)^3}{12x6000}=5.208\times10^6\,E$$

$\alpha_c=0.644$

From Table of IS456-2000:

$\alpha_{c\,min}=L2/L1=6/6=1$

$$L_L/D_L=\frac{5}{(6.25+0.75)}=0.71\sim1$$

$\alpha_{c\,min}=0.7$

α_{cmin} should be < αc min

$\alpha_c = 0.7$

Total moment on slab $= \dfrac{W.L_n}{8} = 273.375$ kNm

$W = w_u \times L_2 \times L_n = 18 \times 6 \times 4.5 = 486$ kN

$L_n = 6 - 1.5 = 4.5$ m

As per Cl.31.4.3.3 of IS456-2000,

Exterior negative design moment is,

$$\frac{0.65}{1+\frac{1}{\alpha_c}} \times M_o = 73.168 \text{ kNm}$$

Where, $\alpha_c = 0.7$.

Interior negative design moment is,

$$0.75 - \frac{0.10}{1+\frac{1}{\alpha_c}} \times M_o$$

Where,

$$\alpha_c = \frac{\sum K_c}{K_s}$$

$= 193.775$ kNm

For column strip (75%),

$= 0.75 \times 193.775 = 145.3309$ kNm

For middle strip (25%),

$= 0.25 \text{ x } 193.775 = 48.44$ kNm

Interior positive design moment is,

$$0.63 - \frac{0.28}{1+\frac{1}{\alpha_c}} \times M_o$$

$= 140.708$ kNm

For column strip (60%),

$$= 0.6 \times 140.708 = 84.43 \text{ kNm}$$

For middle strip (40%),

$$= 0.4 \times 140.708 = 56.28 \text{ kNm}$$

Check for Depth

$$M_{ulim} = 0.138\, f_{ck}.b.d^2$$

$$145.331 \times 10^6 = 0.138 \times 20 \times 3 \times d^2$$

$$d_{cs} = 132.484 \text{ mm} < 275 \text{ mm}$$

$$M_{ms} = 82.4 \text{ kNm}$$

$$d_{ms} = 82.4 \text{ mm} < 175 \text{ mm}$$

Check for Punching Shear

SF = TL – (Load on circular area)

$$= 18 \times 6 \times 6 - \left[\pi(1.775)^2 / 4\right] \times 18 \; [w_n = 18]$$

$$= 648 - 44.54 = 603.45 \text{ kN } [D + d = 1.5 + 0.275 = 1.775 \text{m}]$$

Shear force/m along the perimeter of the circular area $= \dfrac{SF}{\pi(D+d)} = 108.216 \text{ kN/m}$

Nominal shear stress,

$$\varsigma_v = \frac{V_u}{b.d} = \frac{108.216 \text{x} 10^3}{1000 \text{x} 275}$$

$$= 0.394 \text{ N/mm}^2$$

Design shear stress: $\tau_v = K.\tau_c$

Where,

$$K = (0.5 + \beta) \leq 1$$

$$= (0.5 + 6/6) \leq 1$$

$$= 1.5 \leq 1 \rightarrow K = 1$$

$\tau_c' = 0.25\, f_{ck} = 1.118\ \mathrm{N/mm^2}$

$\tau_v < \tau_c$

Section is safe in shear.

A_{st} for exterior negative moment (73.168 kNm), b = 3000mm, d = 275mm,

$$M_u = 0.87.f_y.A_{st}.\left(d - 0.42\left(\frac{0.87.f_y\,A_{st}}{0.36.f_{ck}.b}\right)\right)$$

Or

$K = \dfrac{M_u}{bd^2}$ Take p_t from SP 16.

$$73.168 \times 10^6 = 99.28 \times 10^3 A_{st} - 2.53 A_{st}^{\ 2}$$

$A_{st} = 751.373\ \mathrm{mm^2}$

Required 10mm @ 310mm c/c.

Min A_{st} : 0.12% of c/s $= 0.12/100 \times 3000 \times 275 = 990\ \mathrm{mm^2}$

Provide 10mm @ 230mm c/c.

Similarly, the reinforcement required in CS and MS for –ve and +ve moments are found and listed below:

Location	A_{st} Req.	Min. A_{st}	A_{st} Provided	Rein. Provided
External –ve Moment CS	751	990	990	10 @ 230 c/c
Internal –ve Moment CS	1522	630	791	10 @ 150 c/c
Internal –ve Moment MS	791	630	990	10 @ 290 c/c
+ve Moment CS	869	990	990	10 @ 230 c/c
+ve Moment MS	925	630	925	10 @ 250 c/c

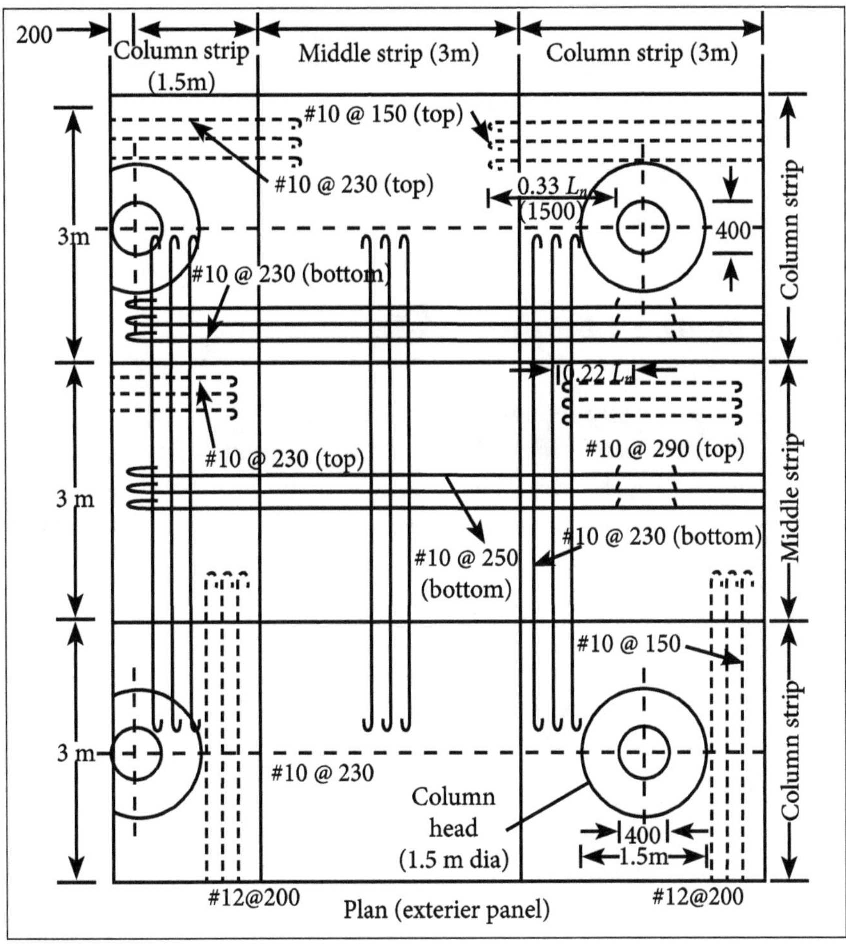

Reinforcement detail.

3

Detailing of Staircases

3.1 Dog Legged and Open Well

Staircase flights are generally designed as slabs spanning between wall supports or landing beams or as cantilevers from a longitudinal inclined beam. The staircase fulfills the function of access between the various floors in the building. Generally, the flight of steps consists of one or more landings provided between the floor levels as shown in the figure below.

The structural components of a flight of stairs comprises of the following elements:

Tread

The horizontal portion of a step was the foot, rest is referred to as tread. 250 to 300 mm is the typical dimensions of a tread.

Riser

Riser is the vertical distance between the adjacent treads or the vertical projection of the step with value of 150 to 190 mm depending upon the type of building. The width of stairs is generally 1 to 1.5 m and in any case not less than 850 mm. Public buildings should be provided with larger widths to facilitate free passage to users and prevent overcrowding.

Going

Going is the horizontal projection (plan) of an inclined flight of steps between the first and last riser. A typical flight comprises two landings and one going as shown in the figure (e).

To break the monotony of climbing, the number of steps in a flight should not generally exceed 10 to 12.

The tread-riser combination can be provided in conjunction with given below:

- Waist slab (a).
- Tread: Riser type (continuous folded plate) (b).
- Isolated cantilever tread slab (c).

- Double cantilever precast tread slab with a central inclined beam (d).

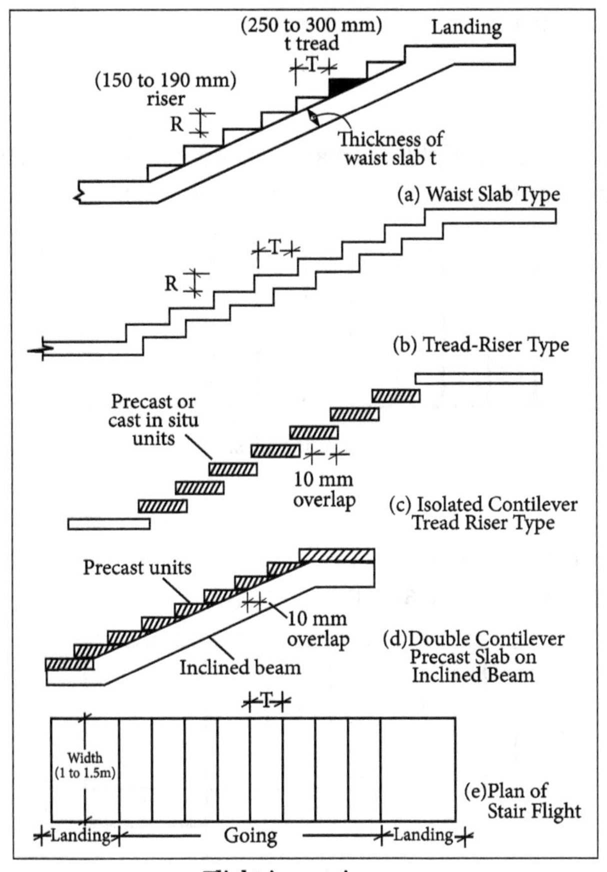

Flight in a staircase.

Types of Staircases

Staircases are broadly classified as:

- Straight stair.
- Quarter turn stair.
- Half turn stair.
- Dog legged stair.
- Open newer stair with quarter space landing.
- Geometrical stairs such as circular stair, spiral stair, etc.

The below are the places where the following staircases can be used:

- Single flight staircase.
- Quarter turn staircase.

- Dog legged staircase.
- Open well staircase.
- Spiral staircase.

Single Flight Staircase: Single flight staircase is used in cellars or attics where the height between floors is small and the frequency of its use is less.

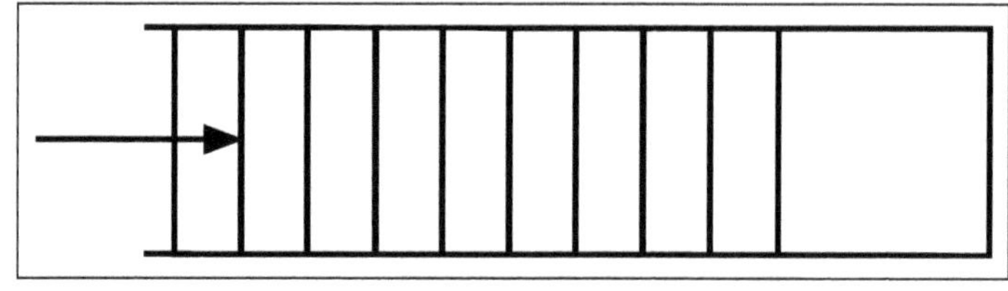

Straight stairs.

Quarter Turn Staircase: Quarter turn staircase flight generally runs adjoining the walls and provides uninterrupted space at the centre of the room. Generally, used in domestic houses where the floor heights are limited to 3m.

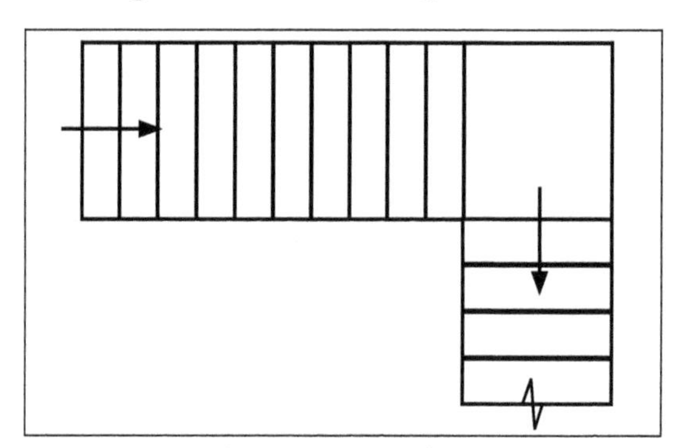

Quarter-turn stairs.

Dog Legged Staircase: Dog legged staircase is generally adopted in economical utilization of the available space.

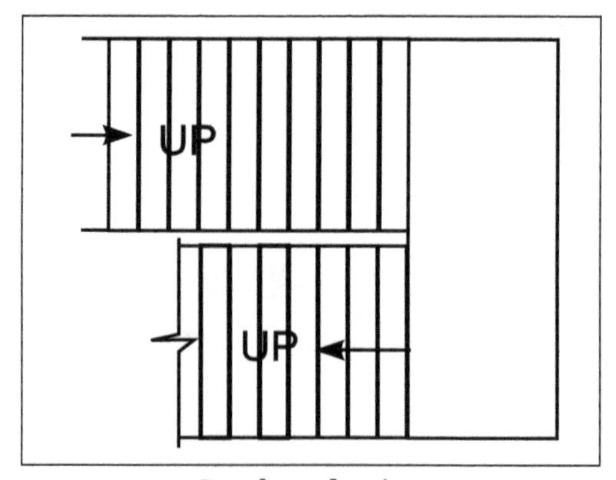

Dog-legged stairs.

Open Well Staircase: Open well staircases are provided in public buildings where large spaces are available.

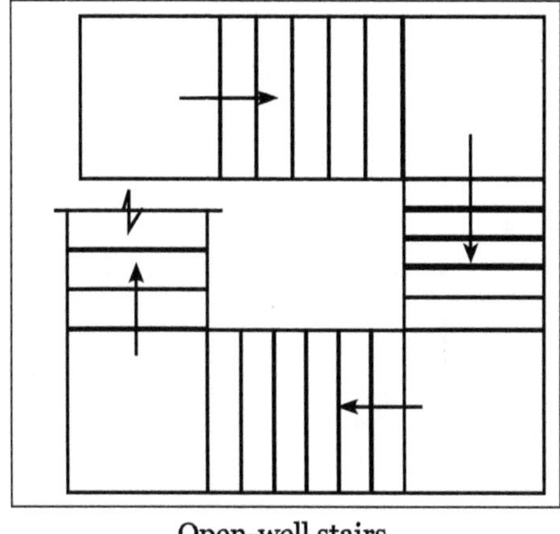

Open-well stairs.

Spiral Staircase: In congested locations where space availability is small, spiral stairs are provided.

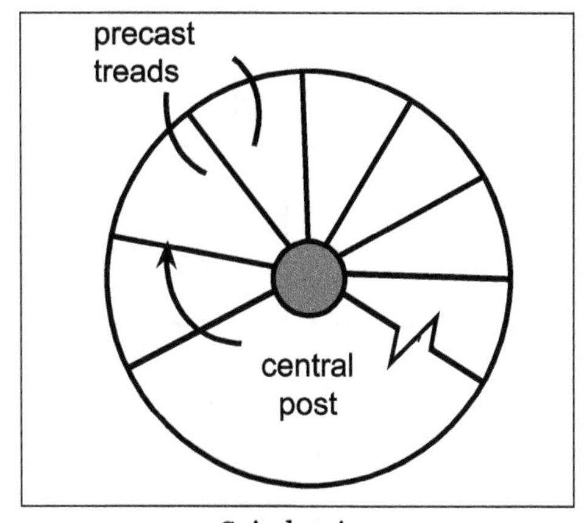

Spiral stairs.

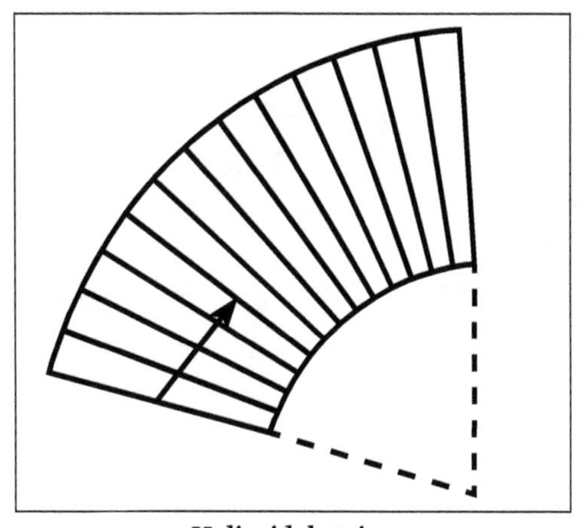

Helicoidal stairs.

Special Support Conditions for Longitudinally Spanning Stair Slabs

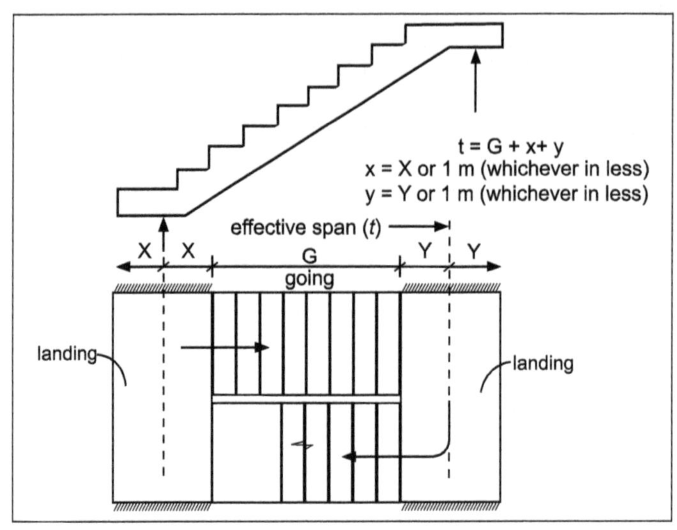

Transversely spanning landings.

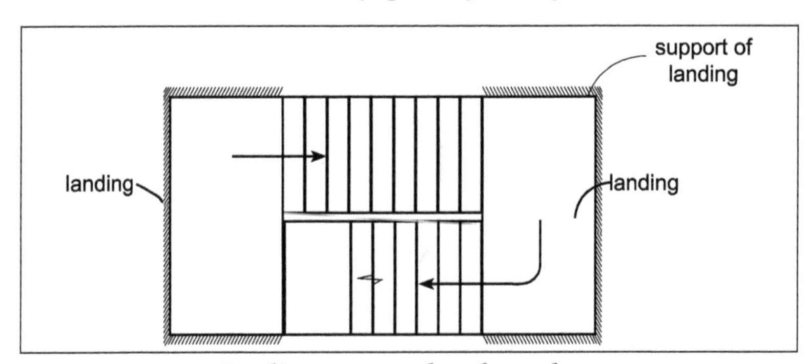

Landings supported on three edge.

Based On Loading and Support Conditions

Spanning Along Transverse Direction

- Cantilever staircase.
- Slab supported between stringer beams.

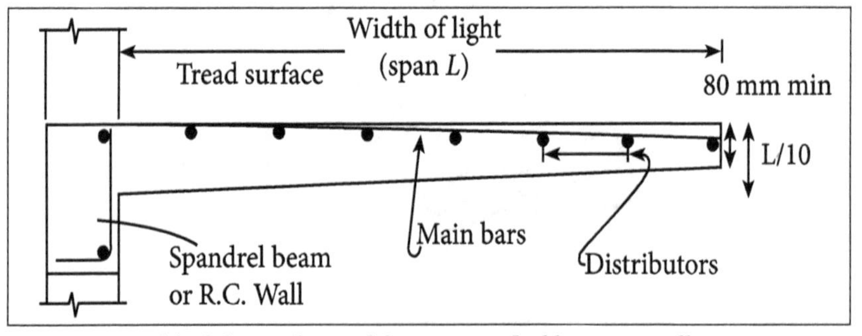

(a) Slab cantilevered from a spandrel beam or wall.

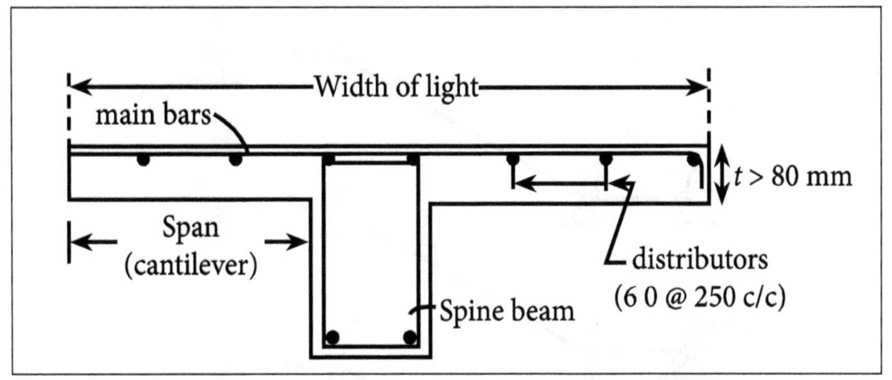

(b) Slab doubly cantilevered from a central spine beam.

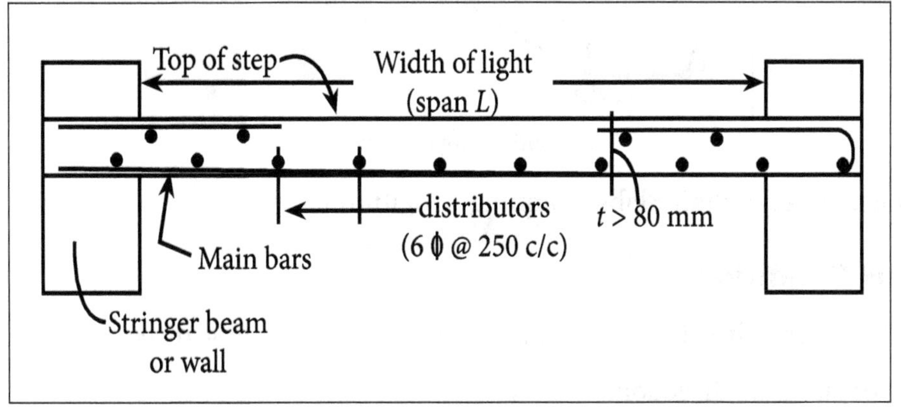

(c) Slab supported between two stringer beam or walls.

Typical Examples of Stair Slabs Spanning Transversely

Spanning Along Longitudinal Direction

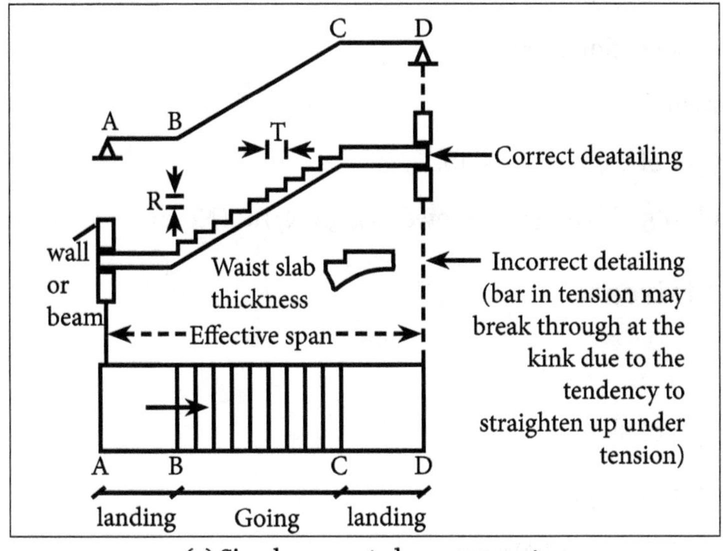

(a) Simply supported arrangement.

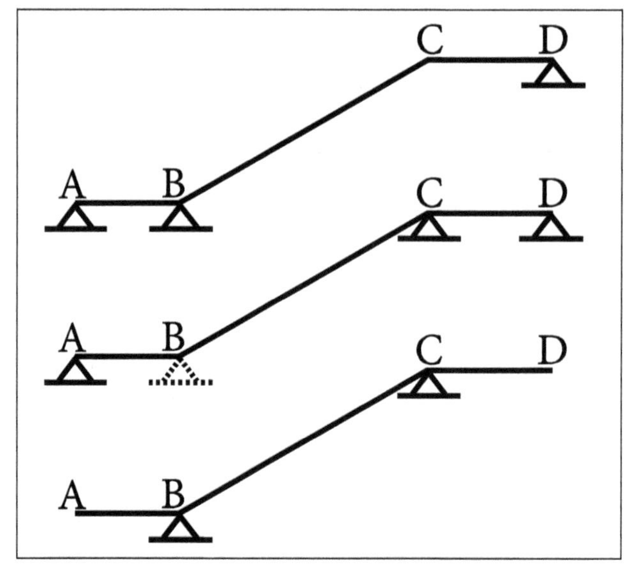

(b) Alternative support arrangement.

Typical Examples of Stair Slabs Spanning Longitudinally:

Support Conditions

- Transverse direction: Stair is spanning along transverse direction.
- Longitudinal direction.
- Stair slab spanning longitudinally and landing slab supported transversely.

In Tread – Riser stair, span by depth ratio is taken as 25 and the loading on the folded slab comprising the tread and riser is idealized as a simply supported slab with loading on landing slab and going similar to a waist like slab.

The loading on folded slab includes:

- Floor finish.
- Self-weight of tread riser slab.
- Live load → 5 kN/m^2 (overcrowded), 3 kN/m^2 (No over-crowding).

Loading on Staircase

Dead Load:

- Self-weight of slab.
- Self-weight of step.
- Tread finish [0.6 – 1 kN/m^2].

Live Load:

- For overcrowding $\rightarrow$ 5 kN/m^2.
- No overcrowding $\rightarrow$ 3 kN/m^2.

RCC Dog-legged Staircase Design

In this type of staircase, the succeeding flights rise in opposite directions. The two flights in plan are not separated by a well. A landing is provided corresponding to the level at which the direction of the flight changes.

Procedure for Dog-legged Staircase Design

Based on the direction along which a stair slab span, the stairs maybe classified into the following two types:

- Stairs spanning horizontally.
- Stairs spanning vertically.

Stairs Spanning Horizontally: These stairs are supported at each side by walls. Stringer beams or at one side by wall or at the other side by a beam.

Loads:

Dead load of a step = $\frac{1}{2} \times T \times R \times 25$

Dead load of waist slab = $b \times t \times 25$

Live load = LL (KN/m^2)

Floor finish = Assume 0.5 KN/m

Stairs Spanning Longitudinally: In this, stairs spanning longitudinally, the beam is supported at top and at the bottom of flights.

Loads:

Self weight of a step = $1 \times R/2 \times 25$

Self weight of waist slab = $1 \times t \times 25$

Self weight of plan $= 1 \times t \times 25\left[\left(R^2 + T^2\right)/T\right]$

Live load $= LL\left(KN/m^2\right)$

Floor finish = assume 0.5 KN/m

For the efficient design of an RCC stair, we have to first analyze the various loads that are going to be imposed on the stair.

The load calculations will help us determine, how much strength is needed to carry the load. The strength bearing capacity of a staircase is determined on the amount of steel and concrete used.

The ratio of steel to concrete has to be as per standards. Steel in the staircase will take the tension imposed on it and the concrete takes up the compression.

These are the essential steps that are to be followed for the RCC Stair Design.

Problems

1. Let us design a dog legged staircase having a waist slab for an office building for the following data:

- Height between floor = 3.2m.
- Riser = 160mm.
- Tread = 270mm.
- Width of flight is equal to the landing width = 1.25m.

$LL = 5\ kN/m^2,\ FF = 0.6\ N/mm^2$

Assume the stairs to be supported on 230mm thick masonry walls at the outer edges of the landing parallel to the risers. Use M20 concrete and Fe415 steel.

Solution:

Given:

Height between floor = 3.2m

Riser = 160mm

Tread = 270mm

Width of flight is equal to the landing width = 1.25m

$LL = 5\ kN/m^2,\ FF = 0.6\ N/mm^2$

Note: Based on riser, number of steps is found. Based on tread, length of staircase is found.

Number of steps = 3.2/0.16 = 20

10 numbers of steps are used for first flight and other 10 to the second flight.

Loading on Going

Self weight of waist slab $= 25 \times 0.283 \times (0.31385/0.270) = 8.22$ kN/m²

Self weight of step $= 25 \times ½ \times 0.16 = 2$ kN/m²

Tread finish $= 0.6$ kN/m²

Live load $= 5$ kN/m²

Total $= 15.82$ kN/m²

Loading on Landing Slab:

Self weight of slab $= 25 \times 0.283 = 7.075$ kN/m²

$$l/d = 20 \rightarrow 5.16/d = 20$$

$$d = 258\,mm$$
$$D = 258 + 20 + 10/2 = 283\,mm$$

Length of inclination of one step,

$$R = 160mm,\ T = 270mm,\ L = 313.85mm$$

Self weight of slab $= 25 \times 0.283 = 7.075$ kN/m²

$$FF = 0.6\ kN/m^2,\ LL = 5\ kN/m^2$$

Total $= 12.675$ kN/m²

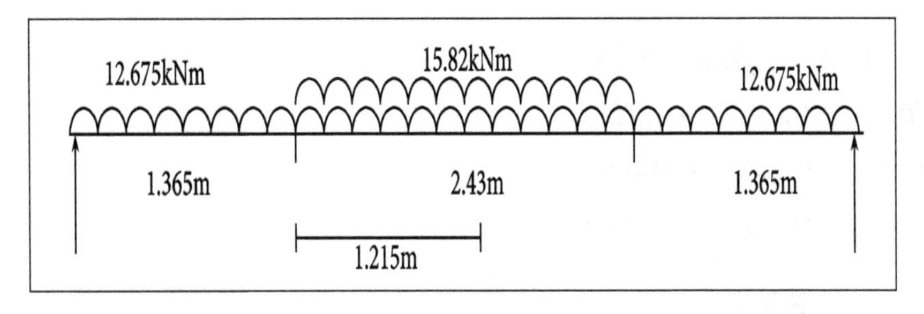

$$(R_A \times 5.16) - (12.675 \times 1.365 \times 4.4775) - (15.82 \times 2.43 \times 2.58)$$
$$-(12.675 \times 1.365 \times 0.6875) = 0$$

$$R_A = 36.54\ kN$$
$$R_B = 36.51\ kN$$

Maximum moment at centre = (36.5 × 2.58) – (12.675 × 1.365 x (0.6825 × 1.215) – (15.82 × 1.2152/2) = 49.66 kNm

Factored moment = 74.49 kNm

$$M_M = 0.87.f_y.A_{st}\left(d - 0.42\left(\frac{0.87.\ f_y\ A_{st}}{0.36.\ f_{ck}.b}\right)\right)$$

$$b = 1000mm,\ d = 258mm$$

$$74.49 \times 10^6 = 93.15 \times 103\ A_{st} - 7.4\ Ast^2$$

$$A_{st} = 868.99\ mm^2$$

Provide 12mm φ @ 130mm c/c.

Distribution Steel:

$$0.15\%\ of\ c/s = 0.15/100 \times 1000 \times 283 = 424.5\ mm^2$$

$$8mm@110mm\ c/c$$

Check for Shear:

$$\zeta_v = V_u/b.d$$

Maximum shear force = [36.5 – (12.675 × 0.258)] × 1.5 = 49.84 kN

$$\zeta_v = 0.193\ N/mm^2$$

Let us find the ζ_c:

$$P_t = 100.\ A_{st}/b.d = 0.336\%$$
$$For\ p_t = 0.25\% \rightarrow 0.36$$
$$For\ p_t = 0.5\% \rightarrow 0.48$$
$$For\ p_t = 0.336\% \rightarrow 0.40$$

$$\zeta_c = 0.40\ N/mm^2$$

Modification factor,

For D = 150 mm, → 1.3
For D = 300 mm, → 1
For D = 283 mm, → 1.03

ζ_c modified $= 1.03 \times 0.40 = 0.412$ N/mm^2

$\zeta_v < \zeta_c$

Safe in shear.

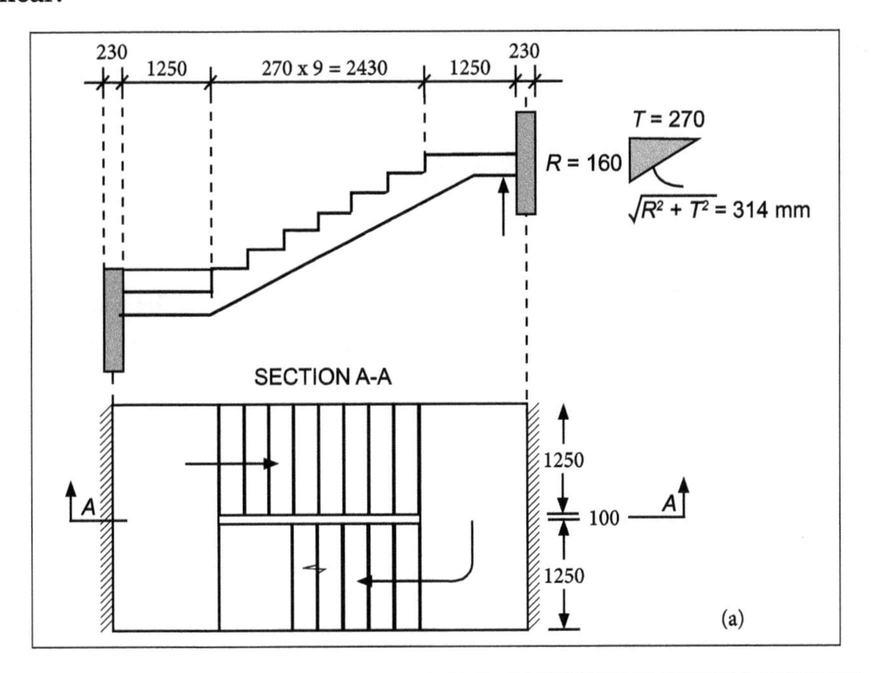

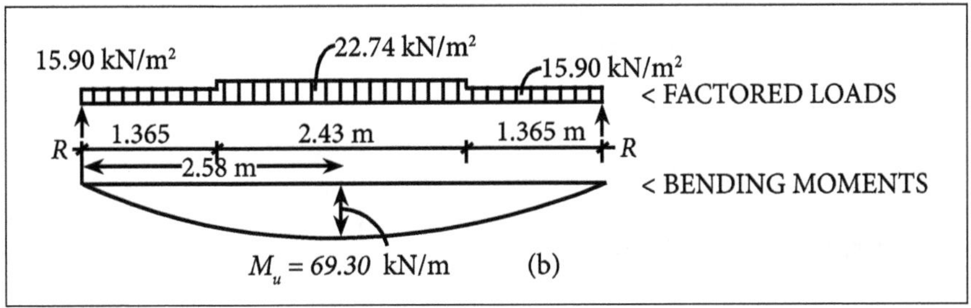

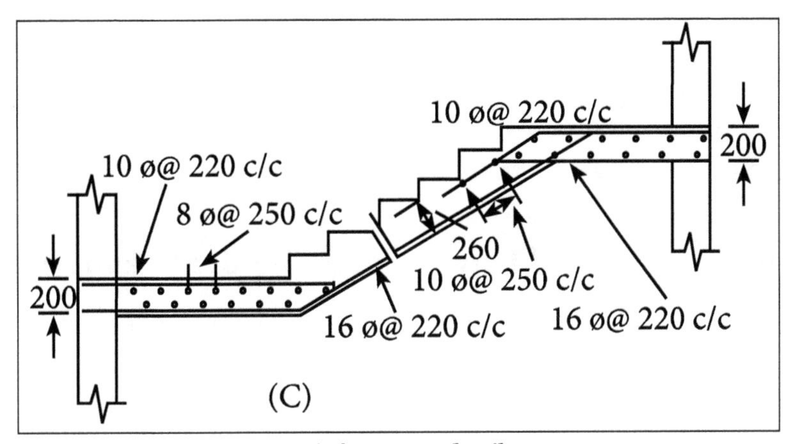

Reinforcement detail.

Tread Riser Staircase

2. Let us design a dog legged staircase having a tread-riser slab for an office building for the following data:

- Height between floor = 3.2m.
- Riser = 160mm.
- Tread = 270mm.
- Width of flight is equal to the landing width = 1.25m.

  $LL = 5\ kN/m^2,\ FF = 0.6\ N/mm^2$

Assume the stairs to be supported on 230mm thick masonry walls supported only on two edges perpendicular to the risers. Use M20 concrete and Fe415 steel. The length of the landing slab is halved while finding the effective length along the longitudinal direction since the staircase is supported only on the landing slab along the transverse direction.

Solution:

Given:

Height between floor = 3.2m.

Riser = 160mm.

Tread = 270mm.

Width of flight is equal to the landing width = 1.25m.

$LL = 5\ kN/m^2,\ FF = 0.6\ N/mm^2$

Effective length = 2.43 + 1.25 = 3.68m.

Assume l/d = 25, 3.68/d = 25,

$$d = 147.2\ mm \sim 150mm$$
$$D = 150 + (20 + 10/2) = 175mm$$

Loading on the Going and Landing Slab: Folder Tread Riser

Self weight of tread riser $= 25 \times (0.27 + 0.16) \times (0.175/0.27) \times 1 = 6.97\ kN/m^2$

Floor finish $=\ 0.6\ kN/m^2$

Live load $=\ 5\ kN/m^2$

Total = 12.57 kN/m²

Considering 1m strip, w = 12.57 kN/m

Loading on Landing Slab:

Self weight of slab = 25 ×0.175 = 4.375 kN/m²

Floor finish = 0.6 kN/m²

Live load = 5 kN/m²

Total = 9.975 kN/m²

Considering 1m strip, w = 9.975 kN/m

50% of load on landing slab is considered along the longitudinal direction.

Along the longitudinal direction, the loading is,

$$(R_A \times 3.68)-(4.99 \times 3.3675)-(12.57 \times 1.84 \times 2.43)-(4.99 \times 0.3125 \times 0.625)=0$$

$$R_A = 18.39 \text{ kN}$$
$$R_B = 18.39 \text{ kN}$$

Moment at centre, (i.e. 1.84m),

$$M_{max} = (18.39 \times 1.84)-(4.99 \times 0.625 \times 1.5275)-(12.57 \times 1.215 \times 0.675)$$

$$= 19.79 \text{ kNm}$$

Factored moment = 29.69 kNm

For b = 1000mm, d = 150mm,

$$K = M_u / b.d^2$$
$$A_{st} = 598.36 \text{ mm}^2/\text{m}$$

Provide 12mm φ, spacing required = 189mm.

Provide 12mm @ 180mm c/c [Main bar as cross links on riser and tread].

Distribution Steel:

$$0.12\% \text{ of c/s} = 0.12/100 \times 1000 \times 175 = 210 \text{ mm}^2$$

Provide 8mm @ 230mm c/c [Distance bar along the width of stair].

Check for Shear:

$$\zeta_v = V_u/b.d, V_u = [18.39 - (4.99 \times 0.15)] \times 1.5 = 26.46 \text{ kN}$$

$$\zeta_v = 0.1764 \text{ N/mm}^2$$

Let us find the ζ_c:

$$100A_{st}/b.d. = 0.3989\%$$
$$\text{For } p_t = 0.25 \rightarrow 0.36$$
$$\text{For } p_t = 0.5 \rightarrow 0.48$$
$$\text{For } p_t = 0.39 \rightarrow (0.1584 + 0.2688) = 0.427$$

Modification Factor (K)

$$\text{For D} = 150\text{mm} \rightarrow 1.3$$
$$\text{For D} = 300\text{mm} \rightarrow 1$$
$$\text{For D} = 175\text{mm} \rightarrow 1.08 + 0.17 = 1.25$$

$$\zeta_{c \text{ modified}} = 0.534 \text{ N/mm}^2$$

$$\zeta_v < \zeta_c$$

Hence safe in shear.

Design of Landing Slab

The landing slab is designed as a simply supported slab which includes the load directly acting on the landing and 50% of the load acting on the going slab.

The loading on the landing is;

- Directly on landing = 9.98 kN/m.
- 50% of load on going slab = (12.57 × 2.43)/2 = 15.27 kN/m.

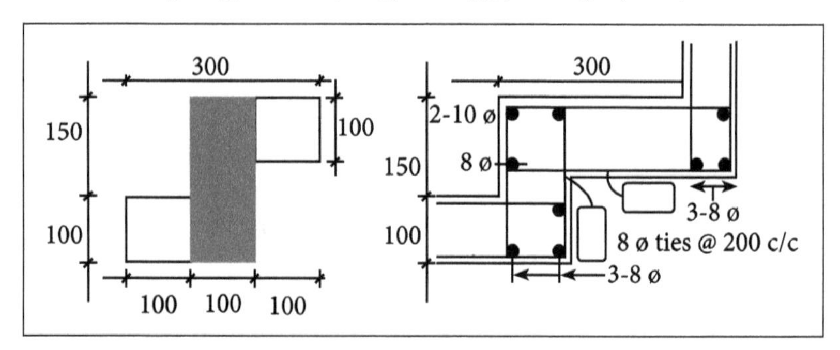

$$w = 25.25 \text{ kN/m}$$

$$l = 2.6\text{m}$$

$$M_u = wl^2/8 \times 1.5 = 32 \text{ kNm}$$

$$b = 1000\text{mm}, \ d = 150\text{mm}$$

$$K = M_u / b.d^2$$

$$A_{st} = 650.19 \text{ mm}^2$$

Providing 12mm φ bar, spacing required = 173.9mm.

Provide 12mm @ 170mm c/c.

Distribution Steel:

0.12% of c/s.

Provide 8mm @ 230mm c/c.

3 numbers of 8mm bars are provided between the cross links as distribution bars. A nominal reinforcement of 10mm @ 200mm c/c is provided at the top of landing slab.

Note: Shear in tread riser slab is negligible. Check for shear is not required.

3. Let us design the open well staircase as shown in the figure (a). The risers are 16 cm, goings are 25 cm and story height is 3.5 m. Goings are provided with 3 cm thick marble finish on cement mortar that weighs 120 kg/m^2, while 2 cm thick plaster is applied to both the risers and bottom surfaces of the slab. The landings are surface finished with terrazzo tiles on sand filling that weighs 160 kg/m^2. The stair is to be designed for a live load of 300 kg/m^2. Use $f'_c = 250 \text{ kg/cm}^2$, $f_y = 4200 \text{ kg/cm}^2$, $\gamma_{plaster} = 2.2 \text{ t/m}^3$.

Solution:

Given:

Flights 1 and 3

Minimum stair thickness required to satisfy deflection requirements is given by,

$$h_{min} = 0.85\left(\frac{420}{20}\right) = 17.85 \text{ cm}$$

Let slab waist t be equal to 18.0 cm.

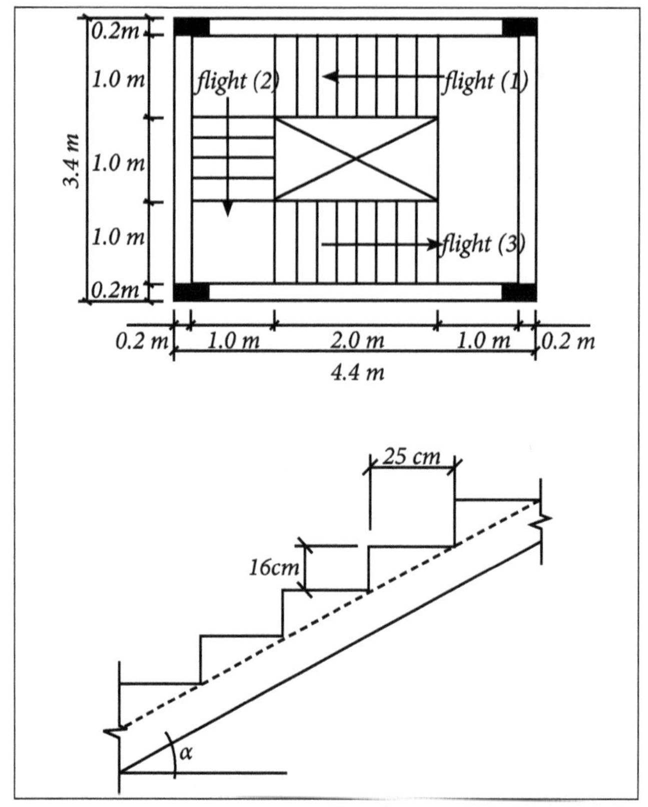

Open-Well stairs.

Loading (Flight)

Dead Load:

$$\text{Own weight of step} = (0.16/2)\ (2.5) = 0.20\ \text{t/m}^2$$

$$\text{Own weight of slab} = (0.18)\ (2.5)/0.8423 = 0.534\ \text{t/m}^2$$

$$\text{Weight of marble finish} = 0.12\ \text{t/m}^2$$

$$\text{Weight of plaster finish} = (0.02)(2.2)/0.8423 = 0.0522\ \text{t/m}^2$$

Live Load:

$$\text{Live load} = 0.3\ \text{t/m}^2$$

Factored Load:

$$w_u = 1.2(0.20 + 0.534 + 0.12 + 0.0522) + 1.6\ (0.3) = 1.57\ \text{t/m}^2$$
$$w_u = 1.57\ (1) = 1.57\ \text{t/m}$$

Loading (Landing)

Dead Load:

$$\text{Own weight of slab} = (0.18)(2.5) = 0.45 \text{ t/m}^2$$

$$\text{Weight of terrazzo finish} = 0.16 \text{ t/m}^2$$

$$\text{Weight of plaster finish} = (0.02)(2.2) = 0.044 \text{ t/m}^2$$

Live Load:

$$\text{Live load} = 0.3 \text{ t/m}^2$$

Factored Load:

$$w_u = 1.2(0.45 + 0.16 + 0.044) + 1.6(0.3) = 1.26 \text{ t/m}^2$$
$$w_{u1} = 1.26(1.5) = 1.89 \text{ t/m}$$
$$w_{u2} = 1.26(0.5) = 0.63 \text{ t/m}$$

Shear Force;

$$V_{u,max} = 3.47 \text{ ton}$$

For φ 14 mm bars:

$$d = 18.0 - 2.0 - 0.7 = 15.3 \text{ cm}$$

$$\phi V_c = 0.75(0.53)\sqrt{250}\,(100)(15.3)/1000 = 9.62 \text{ t} > 3.47 \text{ t}$$

i.e. slab thickness is adequate for resisting beam shear without using shear reinforcement.

Bending Moment:

$$M_{u,max} = 3.28 \text{ t.m as shown in the figure (b).}$$

Flexural reinforcement ratio is given by,

$$\rho = \frac{0.85(250)}{4200}\left[1 - \sqrt{1 - \frac{2.353(10)^5(3.28)}{0.9(100)(15.3)^2(250)}}\right] = 0.00385$$

$$A_s = 0.00385(100)(15.3) = 5.89 \text{ cm}^2$$

Use 6φ12 mm.

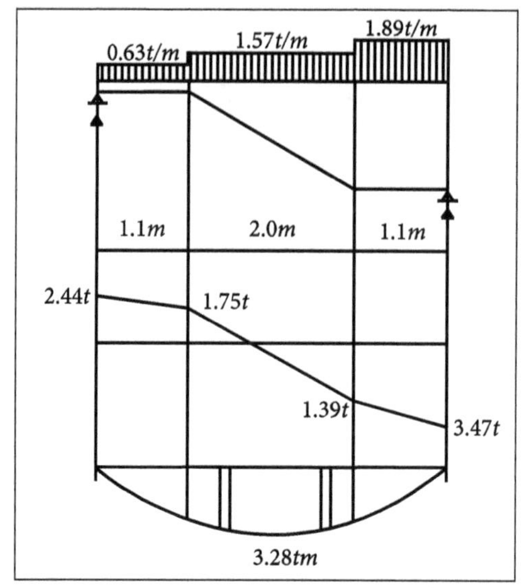

(b) Shear force and bending moment diagrams.

For shrinkage reinforcement, $A_s = 0.0018(100)(18)\ 3.24\ cm^2/m$.

Use 1φ8 mm @ 15 cm in the transverse direction.

Figure (c) shows reinforcement details for flights 1 and 3.

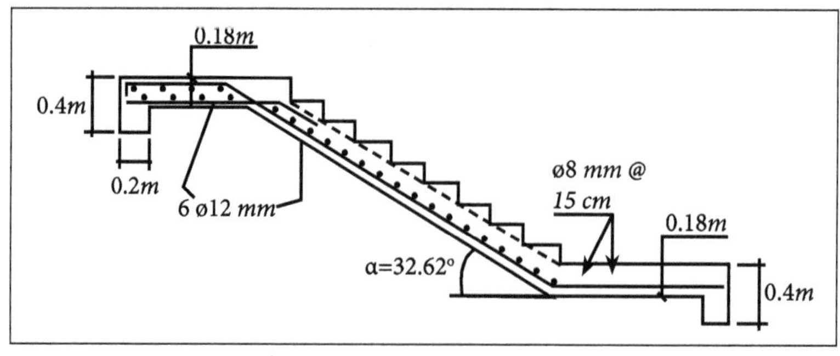

(c) Reinforcement details for flights 1 and 3.

Flight 2

Shear Force:

$$V_{u,max} = 1.672\,t$$

$$d = 18.0 - 3.0 - 0.8 = 14.2\ cm$$

$$\phi V_c = 0.85\ (0.53)\sqrt{250}(100)(14.2)/1000 = 10.11\ t > 1.672\ t$$

i.e. slab thickness is adequate for resisting beam shear without using shear reinforcement.

Bending Moment

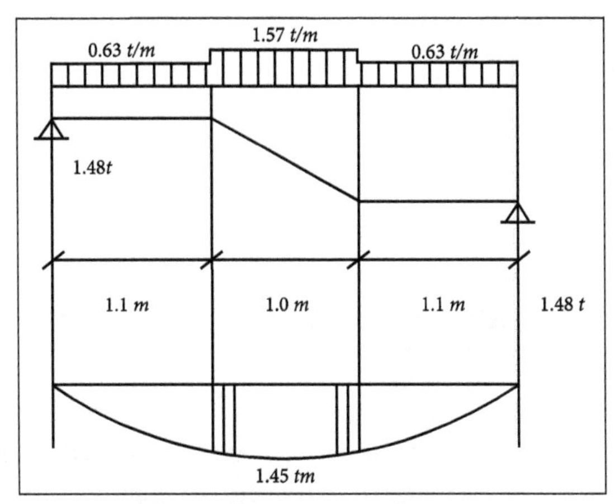

(d) Bending moment diagram.

$M_{u,max} = 1.45$ t.m as shown in the figure (d).

Flexural reinforcement ratio is given by,

$$\rho = \frac{0.85(250)}{4200}\left[1-\sqrt{1-\frac{2.353(10)^5(1.45)}{0.9(100)(14.2)^2(250)}}\right] = 0.00194$$

$$A_s = 0.00194(100)(14.2) = 2.76 \text{ cm}^2$$
$$A_{s,min} = 0.0018(100)\ (18) = 3.24 \text{ cm}^2/\text{m}$$

Use 5φ10 mm.

For shrinkage reinforcement, $A_s = 0.0018(100)\ (18) = 3.24 \text{ cm}^2/\text{m}$.

Use 1φ 8mm@15cm in the transverse direction.

Figure (e) shows reinforcement details for flight 2.

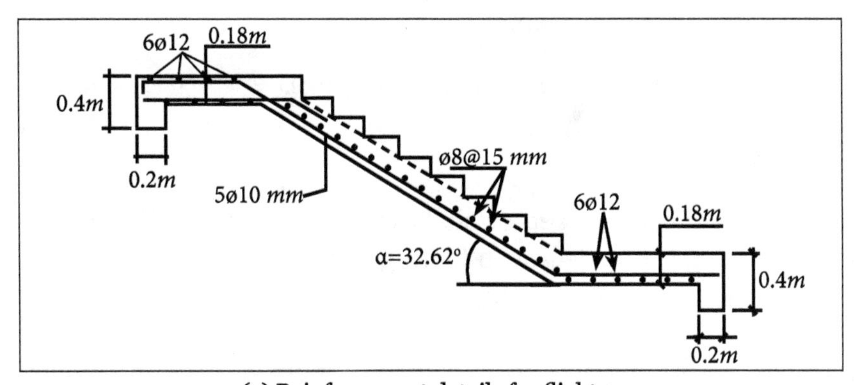

(e) Reinforcement details for flight 2.

4

Detailing of Column Footings

4.1 Column and Footing: Square and Rectangle

Footings are structural elements that transmit column or wall loads to the underlying soil below the structure. Footings are designed to transmit these loads to the soil without exceeding its safe bearing capacity to prevent excessive settlement of the structure to a tolerable limit to minimize differential settlement and to prevent sliding and overturning.

The settlement depends upon the intensity of the load, type of soil and foundation level. Where possibility of differential settlement occurs, the different footings should be designed in such a way to settle independently of each other.

Foundation design involves a soil study to establish the most appropriate type of foundation and a structural design to determine footing dimensions and required amount of reinforcement.

Because compressive strength of the soil is generally much weaker than that of the concrete, the contact area between the soil and the footing is much larger than that of the columns and walls.

The type of footing chosen for a particular structure is affected by the following:

- The bearing capacity of the underlying soil.
- The magnitude of the column loads.
- The position of the water table.
- The depth of foundations of adjacent buildings.

Footing Types

Footings may be classified as deep or shallow. If depth of the footing is equal to or greater than its width, it is called deep footing, otherwise it is called shallow footing. Shallow footings comprise the following types:

Isolated Footings: An isolated footing is used to support the load on a single column. It is usually either square or rectangular in plan. It represents the simplest, most economical type and most widely used footing.

Whenever possible, square footings are provided so as to reduce the bending moments and shearing forces at their critical sections. Isolated footings are used in case of light column loads, when columns are not closely spaced and in case of good homogeneous soil.

Under the effect of upward soil pressure, the footing bends in a dish shaped form. An isolated footing must therefore be provided by two sets of reinforcement bars placed on top of the other near the bottom of the footing.

In case of property line restrictions, footings may be designed for eccentric loading or combined footing is used as an alternative to isolated footing. The figure shows square and rectangular isolated footings.

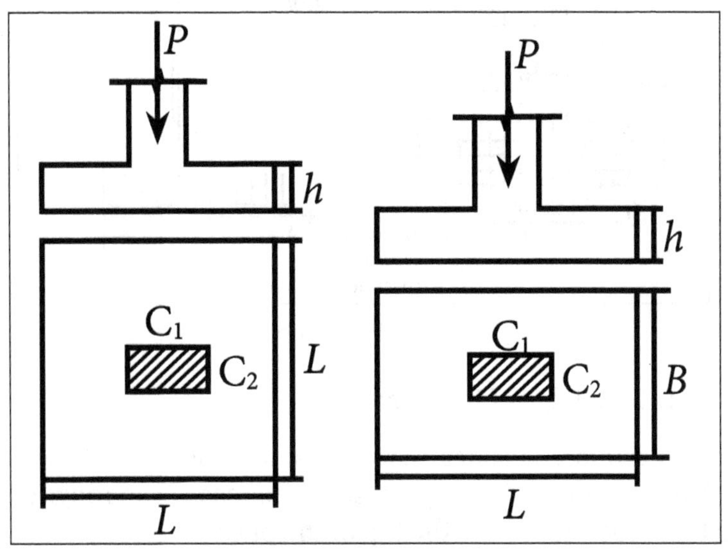

Square and rectangular isolated footings.

Design Guidelines: Specifications for Design of footings as per IS 456: 2000. The important guidelines given in IS 456: 2000 for the isolated footing designs are given below.

Footings is designed to sustain the applied loads, forces and moments and the induced reactions to ensure that any settlement which may occur as uniformly as possible and the safe bearing capacity of the soil is not exceeded.

In sloped or stepped footings the effective cross-section in compression is restricted by the area above the neutral plane. Angle of slope or depth and location of steps is provided such that the requirements of the design are satisfied at every section. Sloped and stepped footings that are designed as a unit can be constructed to assure action as a unit.

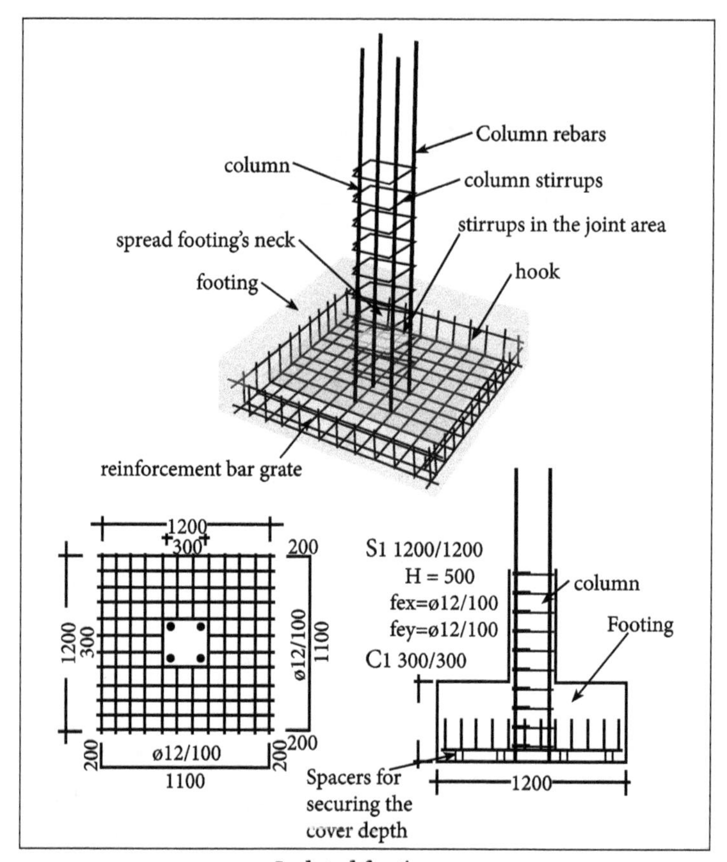

Isolated footings.

General-Thickness at the Edge of Footing:

- In plain and reinforced concrete footings, the thickness at the edge should not be less than 150 mm for footings on soils whereas, for footings on piles it should not less than 300 mm above the top of the piles.
- In the case of plain concrete pedestals, the angle between the plane passing through the bottom edge of the pedestal and the corresponding junction edge of the column with the horizontal plane and pedestal shall be governed by the expression,

$$\tan \alpha <\neq 0.9 * \sqrt{(100 q_o / f_{ck}) + 1}$$

Where,

f_{ck} - Characteristic strength of concrete at 28 days in N/mm².

q_o - Calculated maximum bearing pressure at the base of the pedestal in N/mm².

Moments and Forces:

- In case of footings on piles, computation for moments and shears may be based on the assumption that the reaction from any of the pile is concentrated at the center of the pile.
- For the purpose of computing stresses in footings which supports an octagonal or round concrete column or pedestal, the face of the pedestal or column can be taken as the side of a square which is inscribed within the perimeter of a round or octagonal column or pedestal.

Bending Moment:

- It shall be determined by passing through a vertical plane which entirely extends across the footing and computing the moment of the forces acting over the entire area of the footing on one side of the said plane.
- The greatest bending moment that is to be used in the design of an isolated concrete footing which supports a wall, column or pedestal shall be the moment computed in the manner as follows:
    - At the face of the wall, column or pedestal for the footings supporting a concrete column, pedestal or wall.
    - Halfway between the center-line and the edge of the wall for the footings under masonry walls.
    - Halfway between the face of the pedestal or the column and also at the edge of the gusseted base, for footings under gusseted bases.

Shear and Bond:

- The shear strength of footings is governed by the following two conditions:
    - The footing acting essentially as a wide beam, with a potential diagonal crack extending in a plane across the entire width, the critical section for this condition can be assumed as a vertical section located from the face of the wall, column or pedestal at a distance equal to the effective depth of the footing for footings on piles.
    - Two-way action of the footing, with potential diagonal cracking along the surface of the pyramid or truncated cone around the concentrated load. In this case, the footing shall be designed for shear in accordance with the appropriate provisions.
- In computing the external shear or any section through a footing supported on the piles, the entire reaction from any pile of diameter D_p whose center is

located $D_p/2$ or more outside the section can be assumed as producing shear on the section. The reaction from any pile whose center is located $D_p/2$ or more inside the section can be assumed as producing no shear on the section.

For intermediate positions of the pile center, the portion of the pile reaction should be assumed as producing shear on the section shall be based on the straight line interpolation between the full value at $D_p/2$ outside the section and zero value at $D_p/2$ inside the section.

- The critical section to check the development length in a footing should be assumed at the same plane as those described for the bending moment and also at all the other vertical planes where abrupt changes of section occur. If reinforcement is curtailed, the anchorage requirements must be checked in accordance with 26.2.3 of IS456: 2000.

Tensile Reinforcement:

The total tensile reinforcement at any section will provide a moment of resistance at least equal to the bending moment on the section.

Total tensile reinforcement can be distributed across the corresponding resisting section as given below:

- In one-way reinforced footing, the reinforcement that extends in each direction should be evenly distributed across the full width of the footing.

- In two-way reinforced square footing, the reinforcement extending in both direction shall be evenly distributed across the full width of the footing.

- In two-way reinforced rectangular footing, the reinforcement in the long direction should be uniformly distributed across the full width of the footing.

For reinforcement in the short direction, a central band which is equal to the width of the footing shall be marked along the length of the footing and the portion of the reinforcement determined in accordance with the equation given below shall be uniformly distributed across the central band:

$$\frac{\text{Reinforecement in central band width}}{\text{Total reinforcement in short direction}} = \frac{2}{\beta + 1}$$

Where, â is the ratio between the long side and the short side of the footing.

The remainder of the reinforcement should be distributed uniformly in the outer portions of the footing.

Transfer of Load at the Base of Column:

The compressive stress in concrete at the base of a column or pedestal should be considered as being transferred by the bearing to the top of the supporting footing or pedestal.

The bearing pressure on the loaded area shall not exceed the allowable bearing stress in direct compression multiplied by a value equal to $\frac{\sqrt{A_1}}{\sqrt{A_2}}$ but not greater than 2, where,

A_1 = Supporting area for bearing of footing.

A_2 = Loaded area at the column base.

- When the permissible bearing stress on the concrete in the supporting or the supported member would be exceeded, the reinforcement shall be provided for developing the excess force, either by extending the longitudinal bars into the supporting member or by dowels.
- When transfer of force is accomplished by reinforcement, the development length of the reinforcement shall be sufficient to transfer the tension or compression to the supporting member in accordance with 26.2 of IS456: 2000.
- Extended longitudinal reinforcement or dowels of at least 0.5% of the cross-sectional area of the supported column or pedestal and a minimum of four bars shall be provided. When dowels are used, their diameter shall not exceed the diameter of the column bars by more than 3 mm.
- Column bars of diameters greater than 36 mm, in compression only can be dowelled at the footings with bars of smaller size of the required area. The dowel shall extend into the column, a distance which is equal to the development length of the column bar and into the footing and a distance equal to the development length of the dowel.

Nominal Reinforcement

The minimum reinforcement and the spacing should be as per the requirements of the solid slab. The nominal reinforcement for the concrete sections of thickness greater than 1 m should be 360 mm2 per meter length in each direction on each face.

This provision does not supersede the requirement of the minimum tensile reinforcement based on the depth of the section.

Problem

1. Let us design an isolated footing of uniform thickness of a RC column bearing a

vertical load of 600 kN and having a base of size 500 × 500 mm. The safe bearing capacity of soil may be taken as 120 kN/m². Use M20 concrete and Fe 415 steel.

Solution:

Given:

Base size = 500 × 500 mm

Size of footing:

W = 600 kN;

Self weight of footing @ 10% = 60 kN

Total load = 660 kN

Size of footing $= 660/120 = 5.5 \text{ m}^2$

Since square footing, $B = \sqrt{5.5} = 2.345 \text{ m}^2$

Provide a square footing = 2.4m × 2.4m

Net upward pressure, $p_o = 600/(2.4 \times 2.4) = 104.17 \text{ kN/m}^2$

Design of Section

The maximum BM acts at the face of column,

$$M = M = p_o \frac{B}{8}(B - b)^2 \times 10^6 \text{ N-mm}$$

$$= 104.7 \times (2.4/8)(2.4 - 5)^2 \times 10^6 = 1.128 \times 10^8 \text{ N-mm}$$

$$M_u = 1.5M = 1.5 \times 1.128 \times 10 \text{ N-mm} = 169.2 \text{ kN-m}$$

$$d = \sqrt{\frac{M_u}{R_u B}} = \sqrt{\frac{1.692 \times 10^8}{2.761 \times 2400}} \underline{\Omega}\, 160 \text{ mm}$$

Therefore d = 160 mm;

D = 160 + 60 = 220mm

Depth on the Basis of One-Way Shear

For a one way shear, critical section is located at a distance d(mm) from the face of the column, where shear force V is given by,

$$V = p_o B\{0.5(B - b) - d\} = 104.17 \times 2.4\{0.5 \times (2.4 - 0.5) - 0.001d\} \times 10^3$$
$$= 250008[0.95 - 0.001d]$$

$$V_u = 1.5V = 375012[0.95 - 0.001\,d]$$

$$\hat{o}_c = \frac{V_u}{bd} = \frac{375012(0.95 - 0.001\,d)}{2400\,d}$$

$$= 156.225 / d[0.95 - 0.001d]$$

From table k = 1.16 for D = 220mm.

Also for under-reinforced section with $p_t = 0.3\%$ for M20 concrete, $\hat{o}_c = 0.384\,N / mm_2$.

Hence, design shear stress $= k_c = 0.445\,N / mm_2$

From which we get d = 246.7 = 250 mm.

Depth for Two Way Shear

Take d greater one of the two i.e. 250mm for two-way shear, the section lies at d/2 from the column face all round. The width b_o of the section = b + d = 750mm. Shear force around the section,

$$F = p_o\left[B^2 - b_o^{\ 2}\right] = 541.42\,kN =$$

$$F_u = 1.5F = 1.5 \times 541.424 \times 10^3 = 812136\ N$$

$$\hat{o}_i = \frac{F_u}{4b_o d} = \frac{812.13 \times 10^6}{4 \times 750 \times 250} = 1.083\ N / mm^2$$

Permissible shear stress $= k_s\,\tau_c\tau$

Where, $k_S = (0.5 + \beta_C) = (0.5 + 1)$ with a maximum value 1.

$$\hat{o}_c = 0.25\,\sqrt{f_{ck}} = 1.118\,N / mm^2$$

Permissible shear stress = 1.118 N/mm^2

Hence safe.

Hence, d = 250 mm, using 60 mm as effective cover and keeping D = 330 mm, effective depth = 330-60 = 270 mm in one direction and other direction d = 270- 12 = 258 mm.

Calculation of reinforcement:

$$A_{st} = 1944\ mm^2$$

Using 12 mm bars, spacing required = 138.27 mm

So provide 12 mm @ 125 c/c in each direction.

Development Length

$$L_d = 564 \text{ mm}$$

Provide 60 mm side cover, length of bars available $= 0.5[B - b] - 60 = 890 \text{ mm} > L_d$

So design is safe.

Transfer of load at column base,

$$A_2 = 500 \times 500 = 250000 \text{ mm}^2$$
$$A_1 = [500 + 2(2 \times 330)] = 3312400 \text{ mm}^2$$

$$\sqrt{\frac{A_1}{A_2}} = 3.64$$

Taking,

$$\sqrt{\frac{A_1}{A_2}} = 2$$

Hence permissible bearing stress $= 18 \text{ kN/m}^2$

Actual bearing stress $= 3.6 \text{ N/mm}^2$

Hence safe.

The function of a foundation or a footing is to transmit the load from the structure to the underlying soil. The choice of suitable type of footing mainly depends on the soil condition, the type of superstructure and the depth at which the bearing strata lies.

2. Let us design an isolated footing for a square column, 400mm × 400mm with 12-20 mm diameter longitudinal bars carrying service loads of 1500 kN with M20 and Fe415. The safe bearing capacity of soil is 250 kN/m² at a depth of 1 m below the ground level. Use M-20 and Fe 415.

Solution:

Given:

Step 1:

P = 1500 kN, qc = 250 kN/m² at a depth of 1 m below the ground level.

Assuming the weight of the footing and backfill as 10% of the load, the base area required = 1500(1.1)/250 = 6.6 m². Provide 2.6 m × 2.6 m, area = 6.76 m².

Step 2: Thickness of Footing Slab Based on One-Way Shear.

Factored soil pressure = 1500(1.5)/(2.6)(2.6) = 0.3328 N/mm², say, 0.333 N/mm².

Assuming p = 0.25% in the footing slab, for M20 concrete $c\tau = 0.36\ \text{N/mm}^2$.

$$V_u = 0.36(2600)d \text{ and } V_u(\text{actual}) = 0.333(2600)(1100 - d).$$

From the condition that V_u should be more than or equal to the actual V_u, we have,

$$0.36(2600)d \geq 0.333(2600)\ (1100 - d)$$

So, d ≥ 528.57 mm.

Provide d = 536 mm.

The total depth becomes 536 + 50 + 16 + 8 (with 50 mm cover and diameter of reinforcing bars = 16 mm) = 610 mm.

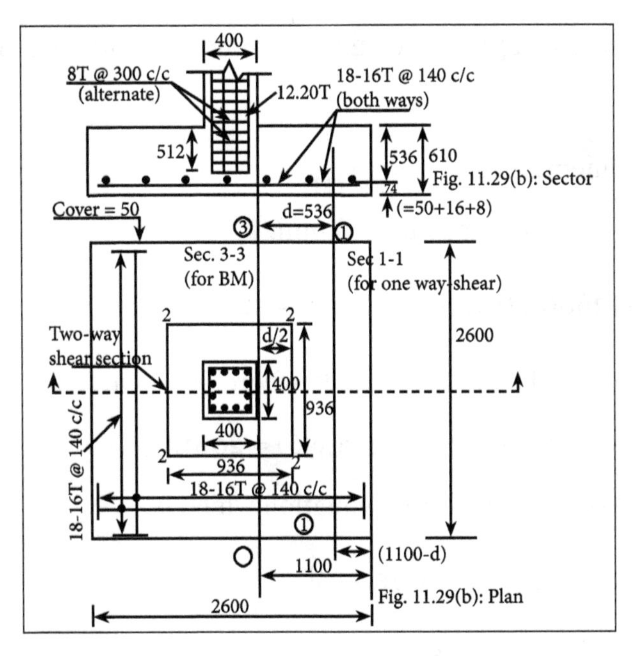

Step 3: Checking for Two-Way Shear.

The critical section is at a distance of d/2 from the periphery of the column.

The factored shear force = 0.333{(2600)² – (400 + d)²}(10)-3 = 1959.34 kN.

Shear resistance is calculated with the shear strength $= k_s\ c\,\tau = k_s(0.25)\ (f_{ck})\ 1/2$

Where $k_s = 0.5 + \beta c$. Here $\beta c = 1.0$, $ks = 1.5 > /1$; so $k_s = 1.0$

This gives shear strength of concrete $= 0.25 \left(f_{ck}\right)1/2 = 1.118\ N/mm^2$

So the shear resistance = (1.118) 4 (936)(536) = 2243.58 kN > 1959.34 kN.

Hence, ok.

Thus the depth of the footing is governed by one-way shear.

Step 4: Gross Bearing Capacity.

Assuming unit weights of concrete and soil as 24 kN/m^3 and 20 kN/m^3 respectively,

Given the service load = 1500 kN

Weight of the footing = 2.6(2.6)(0.61)(24) = 98.967 kN

Weight of soil = 2.6(2.6)(1.0-0.61)(20) = 52.728 kN (Assuming the depth of the footing as 1.0 m).

Total = 1635.2 kN

Gross bearing pressure = 1635.2/(2.6)(2.6) = 241.893 kN/m^2 < 250 kN/m^2.

Hence, ok.

Step 5: Bending Moment.

The critical section is at the face of the column.

$$M_u = 0.333(2600)(1100)(550)\text{Nmm} = 523.809\ \text{kNm}$$

Moment of resistance of the footing = Rbd^2 where, R = 2.76

Moment of resistance = 2.76(2600)(536)(536) = 2061.636 kNm > 523.809 kNm.

Area of steel shall be determined from equation which is:

$$M_u = 0.87\ f_y\ A_{st}\ d\ \left\{1 - \left(A_{st} f_y / f_{ck} bd\right)\right\}$$

Substituting $M_u = 523.809$ kNm, $f_y = 415\ N/mm^2$, $f_{ck} = 20\ N/mm^2$, d = 536 mm, b = 2600 mm, we have:

$$A_{st}^{\ 2} - 67161.44578\ A_{st} + 181.7861758\ \left(105\ \right) = 6$$

Solving, we get $A_{st} = 2825.5805\ mm^2$.

Alternatively, we can use the table of SP-16 to get the A_{st} as explained below,

$$M_u/bd^2 = 523.809(10^6)/(2600)(536)(536) = 0.7013 \text{ N/mm}^2.$$

Table of SP-16 gives p = 0.2034.

$$A_{st} = 0.2034(2600)(536)/100 = 2834.58 \text{ mm}^2.$$

This area is close to the other value = 2825.5805 mm^2.

However one-way shear has been checked assuming p = 0.25%.

So use p = 0.25%.

Accordingly, $A_{st} = 0.0025(2600)(536) = 3484 \text{ mm}^2$.

Provide 18 bars of 16 mm diameter $\left(= 3619 \text{ mm}^2\right)$ both ways.

The spacing of bars = {2600 – 2(50) – 16}/17 = 146.117 mm. The spacing is 140 mm c/c.

The bending moment in the other direction is also the same as it is a square footing. The effective depth, however, is 16 mm more than 536 mm. But the area of steel is not needed to be determined with d = 552 mm as we are providing 0.25% reinforcement based on one-way shear checking.

Step 6: Development length.

$$L_d = f_s \phi / 4(b d \tau) = 0.87(415)(16)/4(1.6)(1.2) = 47(16) = 752 \text{mm}$$

Length available = 1100 – 50 = 1050 mm > 752 mm.

Step 7: Transfer of force at the base of the column.

$$P_u = 1500(1.5) = 2250 \text{ kN}$$

Compressive bearing resistance $= 0.45\, f_{ck} \left(A_1 / A_2\right) 1/2$.

For the column face $A_1 / A_2 = 1$ and for the other face $A_1 / A_2 > 2$ but should be taken as 2.

In any case the column face governs.

Force transferred to the base through column at the interface = 0.45(20)(400)(400) = 1440 kN < 2250 kN.

The balance force 2250 – 1440 = 810 kN has to be transferred by the longitudinal reinforcements, dowels or mechanical connectors.

As it is convenient we propose to continue the longitudinal bars (12-20 mm diameter) into the footing.

The required development length of 12-20 mm diameter bars assuming a stress level of 0.87fy(810/2250) = 129.978 N/mm² is 129.978(20)/4(1.6)(1.2)(1.25) = 270.8 mm.

Here bd τ for M_{20} = 1.2 N/mm² increased factor of 1.6 is due to deformed bars and increased factor of 1.25 is for the compression.

Length available = 610 – 50 – 16 – 16 – 16 = 512 mm > 270.8 mm. Hence, o.k.

The arrangement is shown in the below figure:

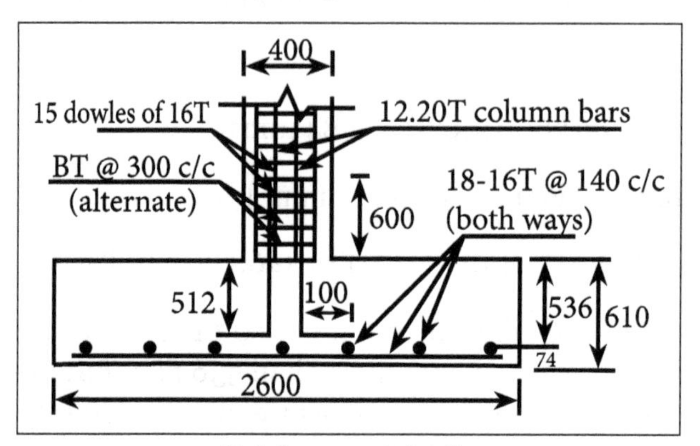

Reinforcement detail.

For the balance force 810 kN the area of dowels = 810000/0.67(415) = 2913.15 mm².

Minimum area = 0.5(400)(400)/100 = 800 mm² < 2913.15 mm².

Therefore number of 16 mm dowels = 2913.15/201 = 15.

The development length of 16 mm dowels in compression = 0.87(415)(16)/4(1.6)(1.2)(1.25) = 601.76 mm.

Available vertical embedment length = 610 – 50 – 16 – 16 – 16 = 512 mm.

So the dowels will be extended by another 100 mm horizontally, as shown in the figure.

Combined Footing

When two or more columns in a straight line are carried out on a single spread footing, it is known as combined footing.

Isolated footings for each column are generally economical. Combined footings are provided only when it is absolutely needed, like:

- When two columns are close to each other, causing overlap of adjacent isolated footings.

- Proximity of building line or sewer or existing building, adjacent to a building column.
- Where soil bearing capacity is low, causing overlap of adjacent isolated footings.

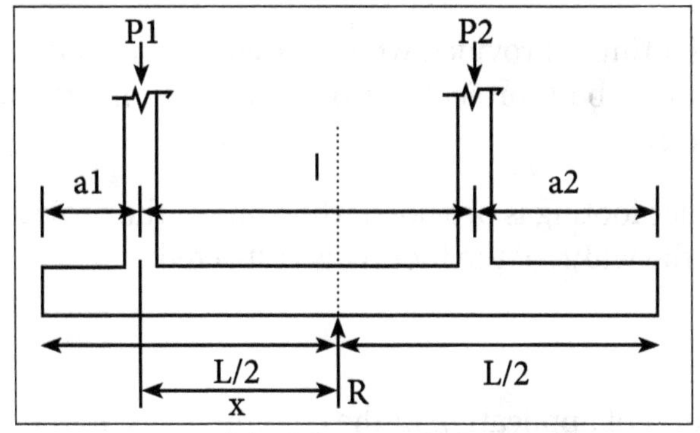

Combined footing with loads.

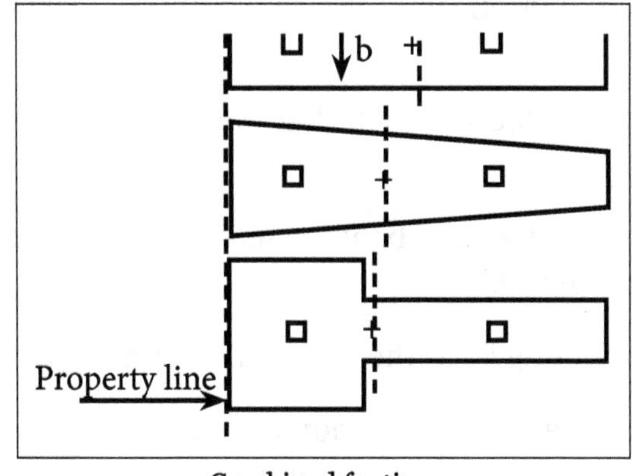

Combined footing.

Types of Combined Footing

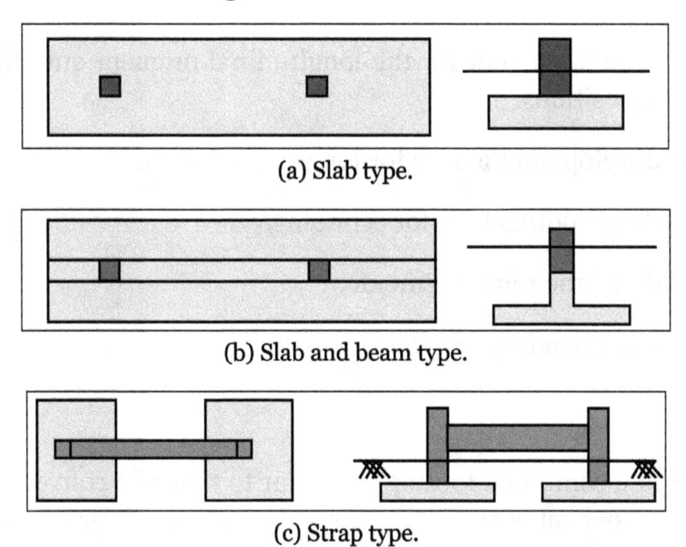
(a) Slab type.

(b) Slab and beam type.

(c) Strap type.

The combined footing may be rectangular, trapezoidal or tee-shaped in plan. The geometric shapes and proportions are so fixed that the centroid of the footing area coincides with the resultant of the column loads. This results in uniform pressure below the entire area of footing:

- Trapezoidal footing is provided when one column load is much greater than the other. As a result, both projections of footing beyond the faces of the columns will be restricted.
- The rectangular footing is provided when one of the projections of the footing is restricted or the width of the footing is restricted.

Design Steps

- Locate the point of application of the column loads on the footing.
- Proportion of the footing such that the resultant of loads passes through the center of footing.
- Compute the area of footing such that the permissible soil pressure is not exceeded.
- Calculate the shear forces and bending moments at the salient points and hence draw the SFD and BMD.
- Fix the depth of footing from the maximum bending moment.
- Calculate the transverse bending moment and design the transverse section for depth and reinforcement. Check for anchorage and shear.
- Check the footing for longitudinal shear and hence design the longitudinal steel.
- Design the reinforcement for the longitudinal moment and place them in the appropriate positions.
- Check the development length for longitudinal steel.
- Curtail the longitudinal bars for economy.
- Draw and detail the reinforcement.
- Prepare the bar bending schedule.

Detailing

Detailing of steel in a combined footing is similar to that of a conventional beam. Detailing requirements of beams and slabs should be followed appropriately.

Types and Proportioning

Mat foundations are primarily shallow foundations. A brief overview of combined footings and the methods used to calculate their dimensions are as follows:

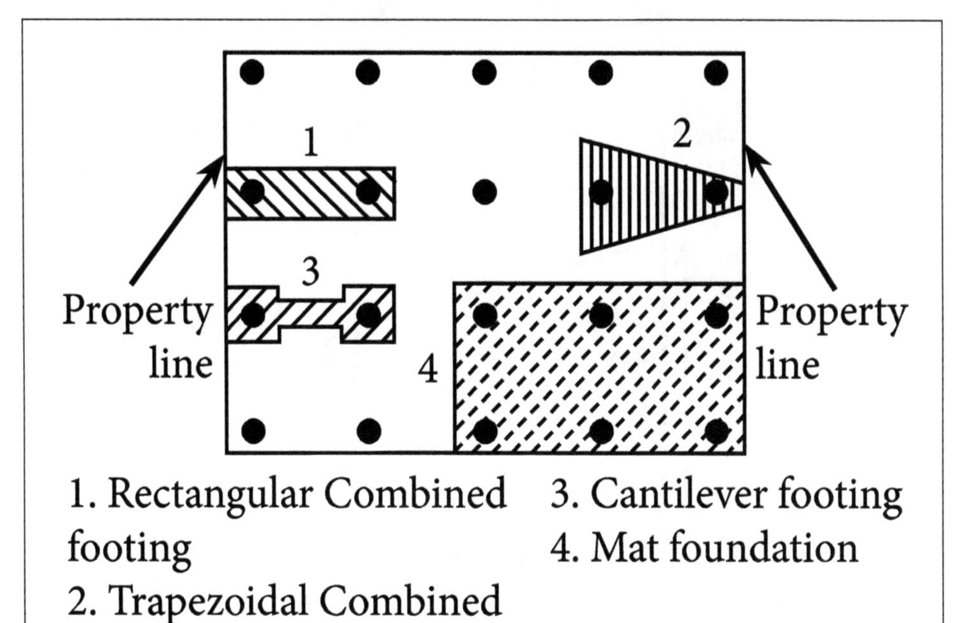

(a) Combined footing.

(b) Rectangular combined footing.

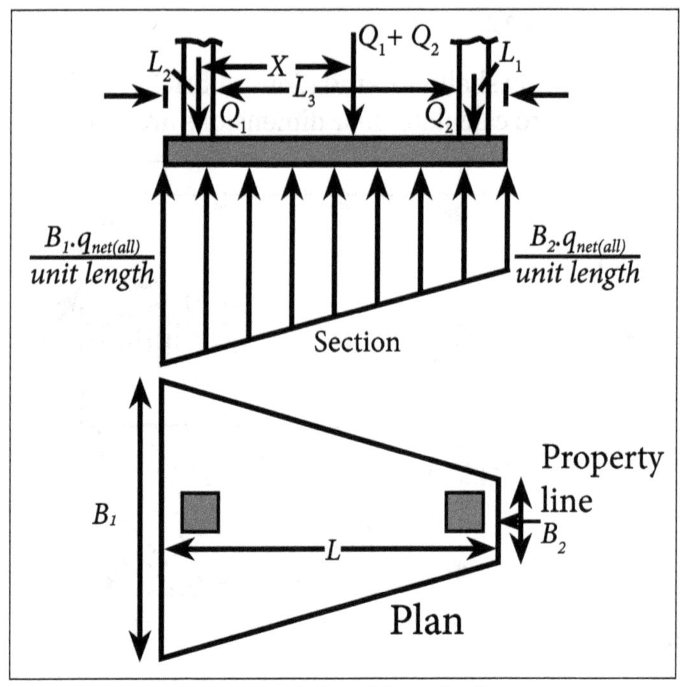

(c) Trapezoidal combined footing.

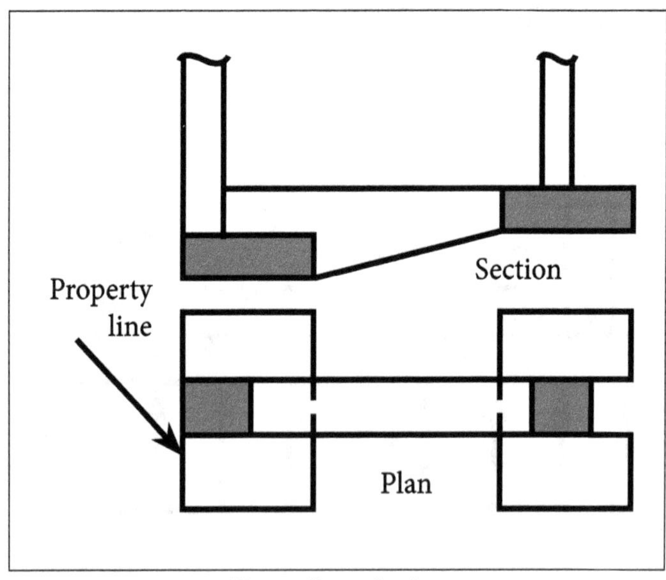

(d) Cantilever footing.

Rectangular Combined Footing

Load to be carried by a column and the soil bearing capacity are such that the standard spread footing design will require an extension of the column foundation beyond the property line. In such a case, two or more columns may be supported on a single rectangular foundation, as shown in the figure.

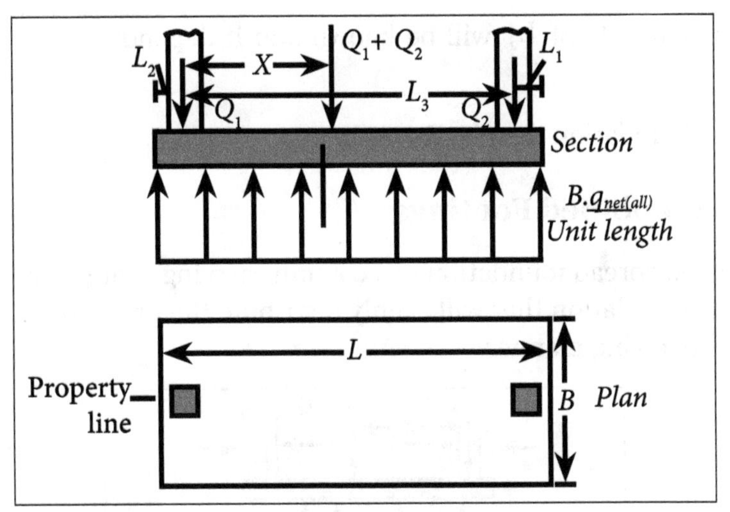

Rectangular combined footing.

If the net allowable soil pressure is known, then the size of the foundation (B x L) can be determined in the following manner:

Determine the area of the foundation, A:

$$A = \frac{Q_1 + Q_2}{q_{all\ (net)}}$$

Where,

$Q_1 + Q_2 =$ Column loads.

$q_{all(net)} =$ Net allowable soil bearing capacity.

Evaluate the location of the resultant of the column loads:

$$X = \frac{Q_2 L_3}{Q_1 + Q_2}$$

For uniform distribution of soil pressure under the foundation, the resultant of the column load must pass through the centroid of the foundation. Thus,

$$L = 2(L_2 + X)$$

Where,

L = Length of the foundation.

Once the length L is found out, the value of L1 can be obtained: $L_1 = L - L_2 - L_3$.

Note that the magnitude of L_2 will be known and it depends on the location of the property line.

The width of the foundation then is: $B = A/L$.

Trapezoidal Combined Footings

Used as an isolated spread foundation of a column carrying a large load where space is tight. Size of the foundation that will evenly distribute the pressure on the soil can be obtained in the following manner.

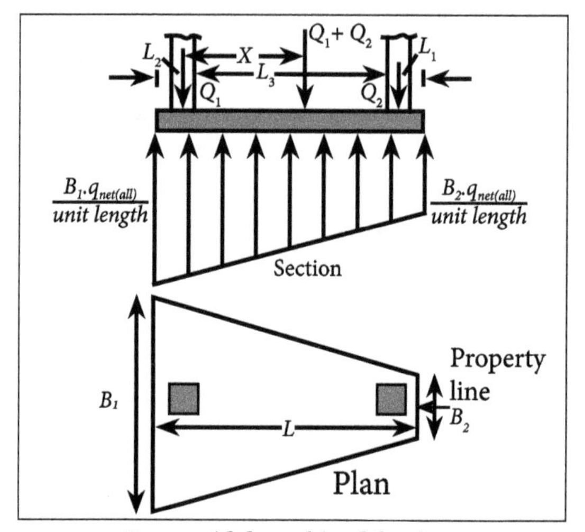

Trapezoidal combined footing.

1. If the net allowable soil pressure is known, determine the area of the foundation:

$$A = \frac{Q_1 + Q_2}{q_{\text{all (net)}}}$$

From above the figure,

$$A = \frac{B_1 + B_2}{2} L$$

2. Determine the location of the resultant for the column loads:

$$X = \frac{Q_2 \, L_3}{Q_1 + Q_2}$$

3. From the property of Trapezoid:

$$X + L_2 = \left(\frac{B_1 + 2B_2}{B_1 + B_2} \right) \frac{L}{3}$$

Cantilever Footing

This type of footing construction uses a strap beam to connect an eccentrically loaded column foundation to the foundation of an interior column.

Used in place of trapezoidal or rectangular combined footings when the distance between the columns are large and the allowable soil bearing capacity is high.

Mat foundation

- It is a combined footing that may cover the entire area under a structure supporting several walls and columns.
- It is sometimes preferred for soils that have low load-bearing capacities but that will have to support high column and wall loads.
- Under certain conditions, spread footings would have to cover more than half of the building area and in such a case mat foundations might be more economical.

1. Let us design an appropriate footing/footings to support two columns A and B spaced at distance 2.1 m center-to-center, as shown in figure is 20 cm × 30 cm and carries a dead load of 20 tons and a live load of 10 tons. Column B is 20 cm × 40 cm in cross section but carries a dead load of 30 tons and a live load of 15 tons. Width of footing is not to exceed 1.0m and there is no property line restriction. Use $f'_c = 300 kg/cm^2$, $f_y = 300 kg/cm^2$, $q_{all}(gross) = 2.0\ kg/cm^2$, $\gamma_{soil} = 1.7 t/m^3$ and $D_f = 2.0$ m.

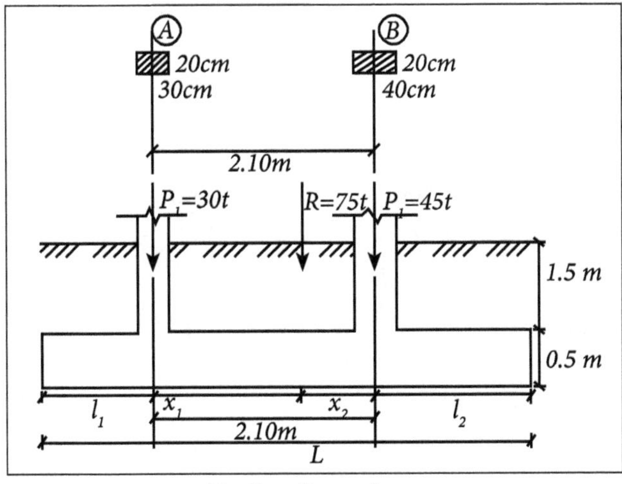

Footing dimensions.

Solution:

Given:

Distance 2.1 m

Dead load = 20 tons

Live load = 10 tons

Column B Carries

Dead load = 30 tons

Live load = 15 tons

$$f'_c = 300kg/cm^2,\ f_y = 4200kg/cm^2,\ q_{all}(gross) = 2.0\ kg/cm^2,$$

$$\Upsilon_{soil} = 1.7t/m^3 \text{ and } D_f = 2.0\ m$$

Select a trial footing depth: Assume that the footing is 50 cm thick.

Establish the required base area of the footing:

$$q_{all}(net) = 20 - 1.5(1.7) - 0.5(2.5) = 16.2t/m^2$$

If isolated footings are to be used,

$$A_{1req} = \frac{P_A}{q_{all(net)}} = \frac{30}{16.2} = 1.85\ m^2$$

B = 1m and L = 1.85 m.

$$A_{2req} = \frac{P_B}{q_{all(net)}} = \frac{45}{16.2} = 2.78\ m^2$$

B = 1m and L = 1.85 m.

Over lapping of the two footings occurs, ruling out isolated footings as appropriate of footing types. A combined footing will be used instead.

$$A_{req} = \frac{P_A + P_B}{q_{all}(net)}$$

$$A_{req} = \frac{20+10+30+15}{16.2} = 4.63\ m^2$$

To locate the resultant of the column forces, replace the column load system with a resultant force system or,

$$P_B(2.1) = R\,(x_1) \text{ or } 45(2.1) = (30+45)(x_1) \text{ and } x_1 = 1.26\ m$$

$$X_2 = 2.1 - 1.26 = 0.84\ m$$

$$L_1 + x_1 = l_2 + x_2 \text{ or}$$

$$l_2 - l_1 = x_1 - x_2 = 0.42 \text{ m}$$

Let B = 1m, length of the footing L = 4.63/1.0 = 4.63m, taken as 4.65m,

$$l_1 + l_2 + 2.1 = 4.65 \text{ m}$$

$$l_1 + l_2 = 2.553$$

Adding (a) and (b) gives, $l_2 = 1.485$ m and $l_1 = 1.065$ m.

Evaluate the net factored soil pressure:

$$P_{Au} = 1.2(20) + 1.6\ (10) = 40 \text{ tons}$$
$$P_{Bu} = 1.2(30) + 1.6(15) = 60 \text{ tons}$$

$$q_u(\text{net}) = \frac{P_{Au} + P_{Bu}}{L\ B} = \frac{40 + 60}{4.65(1.0)} = 21.51 \text{ t/m}^2$$

Check footing thickness for punching shear:

Effective depth d = 50 – 7.5 – 0.80 = 41.7 cm (lower layer).

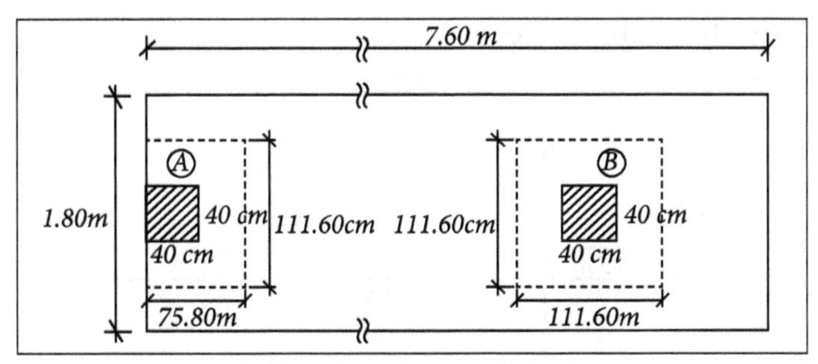

Critical sections for punching shear.

Column A: The factored shear force $V_u = 40 - (21.51)(0.717)(0.617) = 30.48$ tons

$$b_o = 2(71.7 + 61.7) = 266.8 \text{ cm.}$$

ϕV_c is the smallest.

$$\Phi V_c = 0.53\ \Phi\ \sqrt{f'_c}\left(1 + \frac{2}{\hat{a}}\right)\lambda\, b_o\, d$$

$$= 0.53(0.75)\sqrt{300}\left(1 + \frac{2}{30/20}\right)(266.8)(41.7)/1000 = 178.73 \text{ tons}$$

$$\Phi V_c = \lambda \Phi \sqrt{f'_c}\, b_o d$$

$$= 0.75\sqrt{300}\,(266.8)(41.7)/1000 = 144.53 \text{ tons}$$

$$\Phi V_c = 0.27\Phi\left(\frac{\alpha_s d}{b_o} + 2\right)\lambda\sqrt{f'_c}\, b_o d$$

$$= 0.27(0.75)\left(\frac{40(41.7)}{266.8} + 2\right)\sqrt{300}\,(266.8)(41.7)/1000 = 322 \text{ tons}$$

$$\Phi V_c = 144.53 \text{ tons} > 30.48 \text{ tons.}$$

Column B: Effective depth d = 50 – 7.5 – 0.80 = 41.7 cm (lower layer).

The factored shear force,

$$V_u = 60 - (21.51)(0.817)(0.617) = 49.16 \text{ tons}$$

$$b_o = 2(81.7 + 61.7) = 286.8 \text{ cm.}$$

Draw S.F.D and B.M.D for footing.

S.F.D and B.M.D are shown in figure:

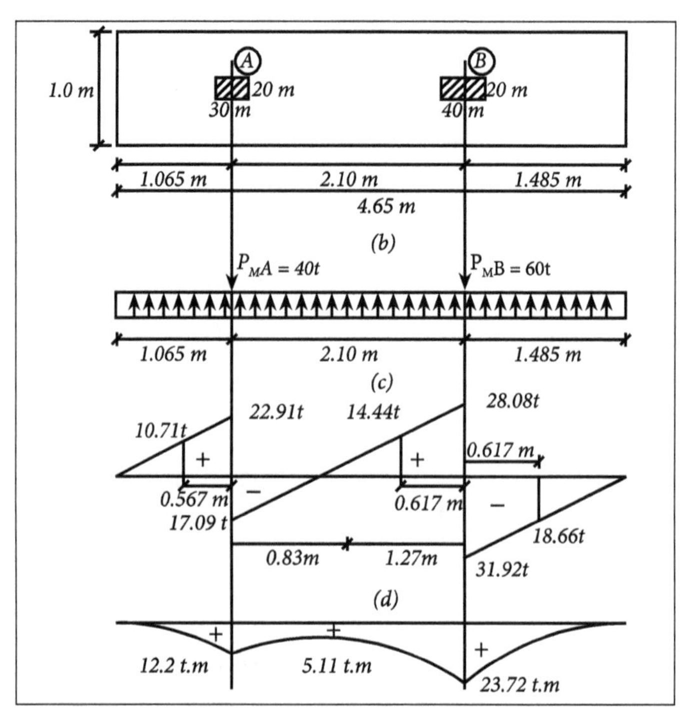

Check footing thickness for beam shear:

$$\Phi V_c = 0.75(0.53)\sqrt{300}(100)(41.7)/1000 = 28.71 \text{ tons}$$

Maximum shear force Vu is located at a distance d from the exterior face of column B,

$$V_u = 18.66 \text{ tons} < 28.71 \text{ tons O.K.}$$

Compute the area of flexural reinforcement:

Bottom longitudinal reinforcement:

$$\rho = \frac{0.85(300)}{4200}\left[1-\sqrt{1-\frac{2.353(10)^5(23.72)}{0.9(100)(41.7)^2(300)}}\right] = 0.0037$$

$$A_s = 0.0037(100)(41.7) = 15.43\text{cm}^2, \text{ use } 8\ \phi 16 \text{ mm}$$

Transverse reinforcement:

Effective depth d = 50 – 7.5 – 1.6 - 0.80 = 40.1 cm (upper layer).

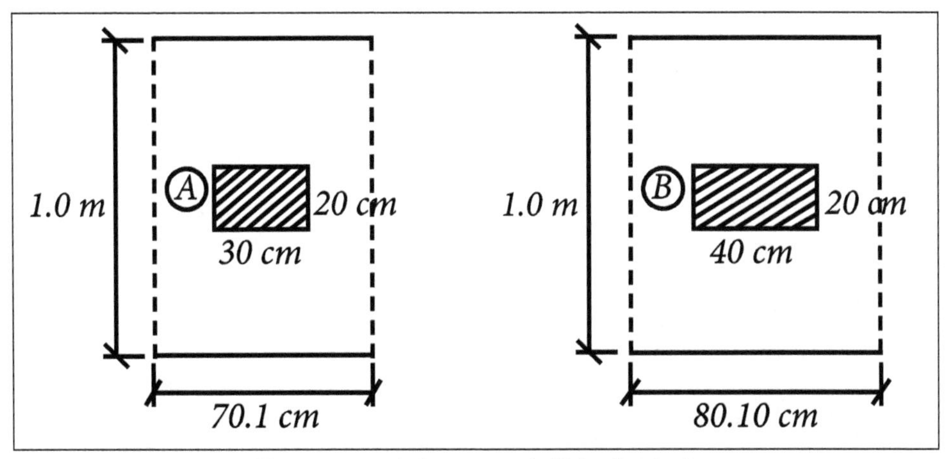

Transverse reinforcement (width of strips).

Under Column A:

$$M_{uA} = \frac{40}{(0.701)(1.0)}\frac{(0.701)}{2}\left(\frac{1-0.2}{2}\right)^2 = 3.2 \text{ t.m}$$

$$\rho = \frac{0.85(300)}{4200}\left[1-\sqrt{1-\frac{2.353(10)^5(3.2)}{0.9(70.1)(40.1)^2(300)}}\right] = 0.00076$$

Under Column A:

so use $\tilde{n} = \tilde{n}_{min} = 0.0018$

$A_s = 0.0018(70.1)(50) = 6.31\ cm^2$ use 5ϕ 14 mm

Under Column B: Shrinkage Reinforcement in the short direction (bottom side).

$$A_s = 0.0018(465 - 80.1 - 70.1)(50) = 28.33\ cm^2 \text{ use } 26\phi \text{ 12 mm}$$

Check bearing strength of column and footing concrete:

$$\phi P_{An} = [0.65(0.85)(300)(20)(30)]/1000 = 199.45 \text{ tons} > 40 \text{ tons}$$

Use minimum dowel reinforcement, $A_s = 0.005(20)(30) = 3.0\ cm^2$

$$\phi P_{Bn} = [0.65(0.85)(300)(20)(40)]/1000 = 132.6 \text{ tons} > 60 \text{ tons}$$

Use minimum dowel reinforcement, $A_s = 0.005(20)(40) = 4.0\ cm^2$

Check for anchorage of the reinforcement:

Bottom longitudinal reinforcement (ϕ 16 mm)

$$\Psi_t = \Psi_e = \lambda = 1 \text{ and } \Psi_s = 0.8]$$

c_b is the smaller of:

$$7.5 + 0.8 = 8.3 \text{ cm, or } \frac{100 - 15 - 1.6}{8(2)} = 5.21 \text{ cm, i.e., } c_b = 5.21 \text{ cm}$$

$$\frac{c_b + K_{tr}}{d_b} = \frac{5.21 + 0}{1.6} = 3.26 > 2.5, \text{ take it equal to } 2.5$$

$$l_d = \frac{1.6(0.8)(4200)}{3.5(2.5)\sqrt{300}} = 35.47 \text{ cm}$$

Available length = 106.5 −7.5 = 99 cm > 35.47 cm.

Similarly, bottom transverse reinforcement is to be checked for anchorage and will be found satisfactory.

Prepare neat design drawings showing footing dimensions and provide reinforcement:

Detailed design drawings are shown in the figure,

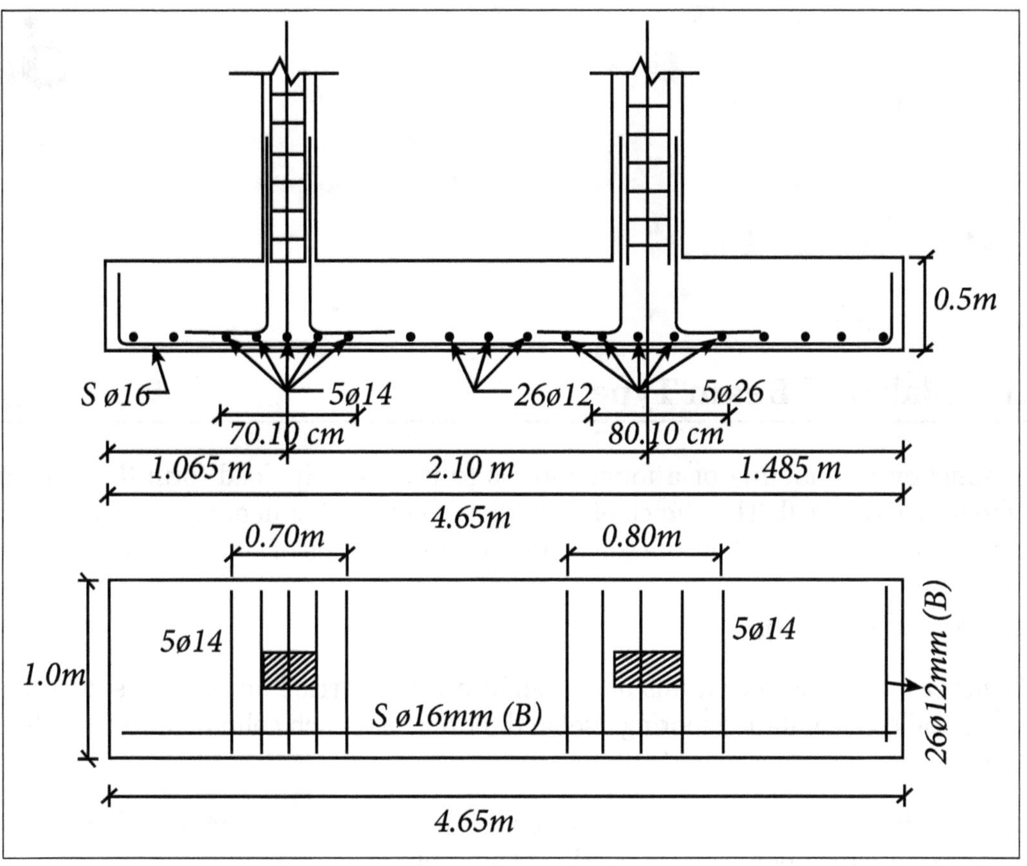

Reinforcement detail.

5

Design and Detailing of Rectangular Combined Footing

5.1 Slab and Beam Type

The function of a footing or a foundation is to transmit the load form the structure to the underlying soil. The choice of suitable type of footing depends on the depth at which the bearing strata lies, the soil condition and the type of superstructure.

Combined Footing

Whenever two or more columns in a straight line are carried on a single spread footing, it is called a combined footing. Isolated footings for each column are generally the economical.

Combined footings are provided only when it is absolutely necessary as, when two columns are close together, causing overlap of adjacent isolated footings.

Where soil bearing capacity is low, causing overlap of adjacent isolated footings.

Proximity of building line or existing building or sewer, adjacent to a building column.

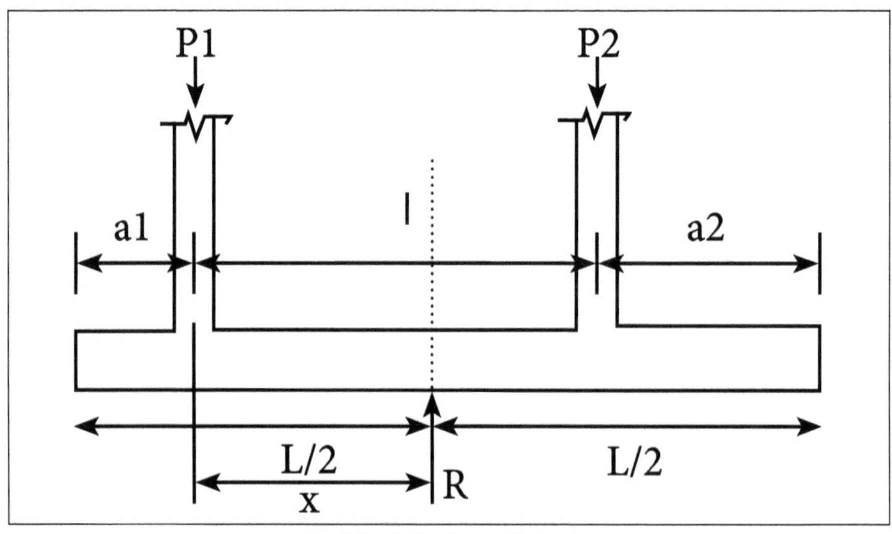

Combined footing with loads.

Types of Combined Footings

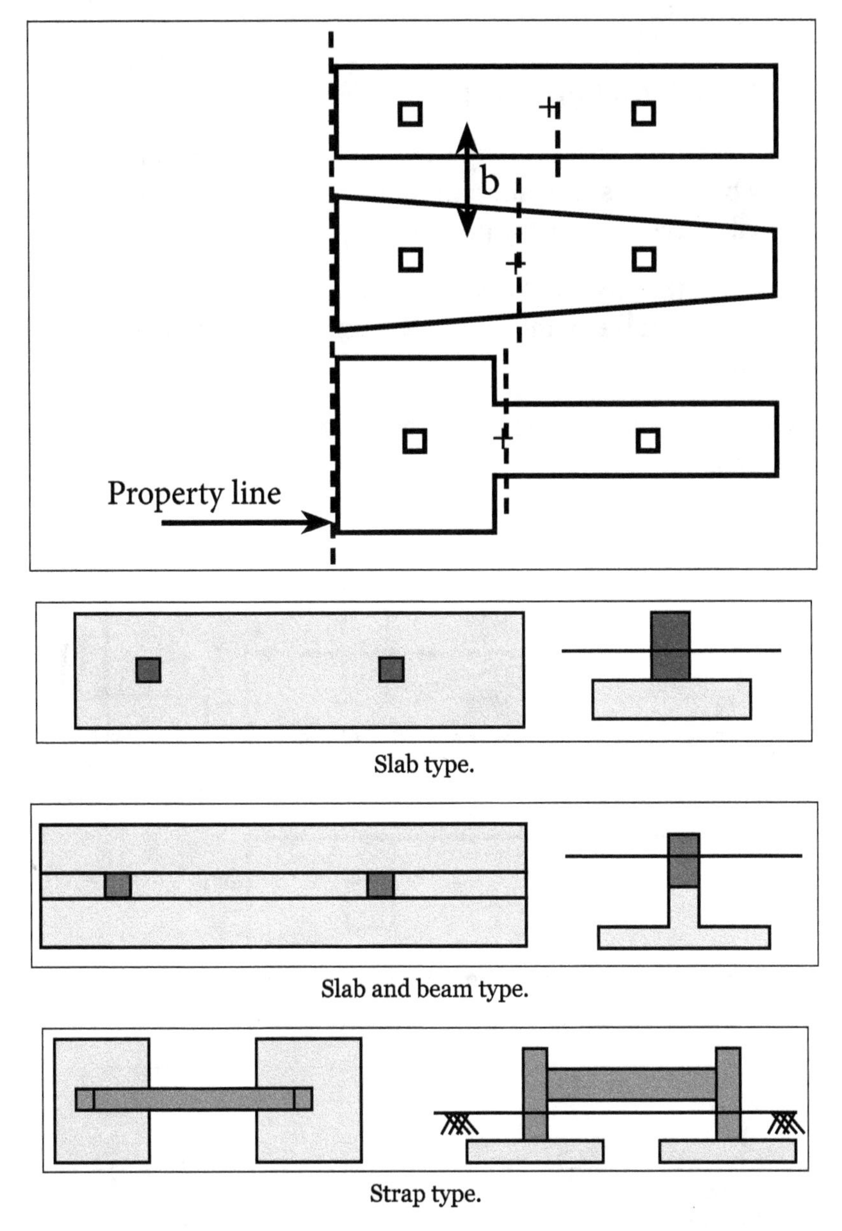

Slab type.

Slab and beam type.

Strap type.

The combined footing may be rectangular, trapezoidal or Tee-shaped in plan.

The geometric proportions and shape are so fixed that the centroid of the footing area coincides with the resultant of the column loads. This results in uniform pressure below the entire area of footing.

Trapezoidal footing is provided when one column load is much more than the other. As a result, the both projections of footing beyond the faces of the columns will be restricted.

Rectangular footing is provided when one of the projections of the footing is restricted or the width of the footing is restricted.

Rectangular Combined Footing

Longitudinally, the footing acts as an upward loaded beam spanning between columns and cantilevering beyond. Using statics, the shear force and bending moment diagrams in the longitudinal direction are drawn.

Moment is checked at the faces of the column. Shear force is critical at distance 'd' from the faces of columns or at the point of contra flexure. Two-way shear is checked under the heavier column.

The footing is also subjected to transverse bending and this bending is spread over a transverse strip near the column.

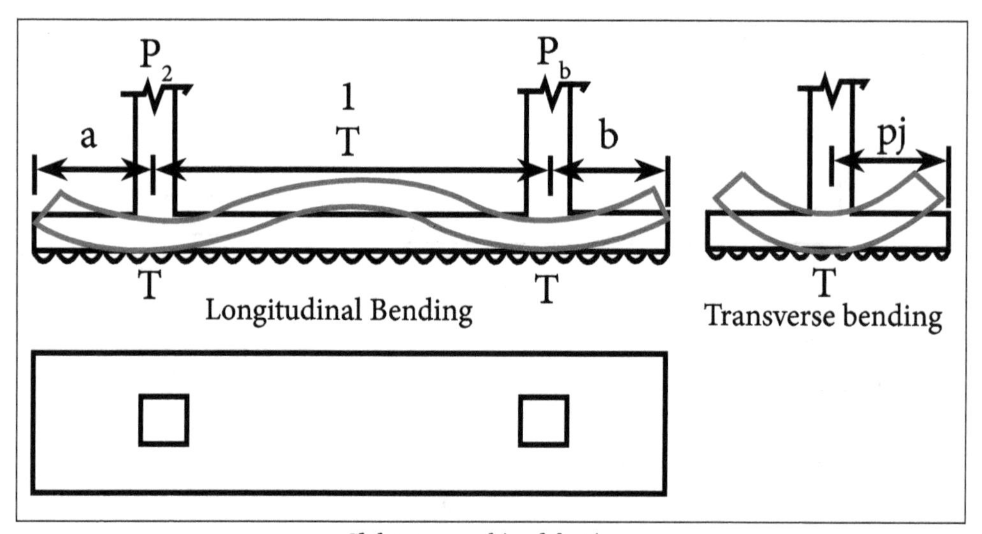

Slab type combined footing.

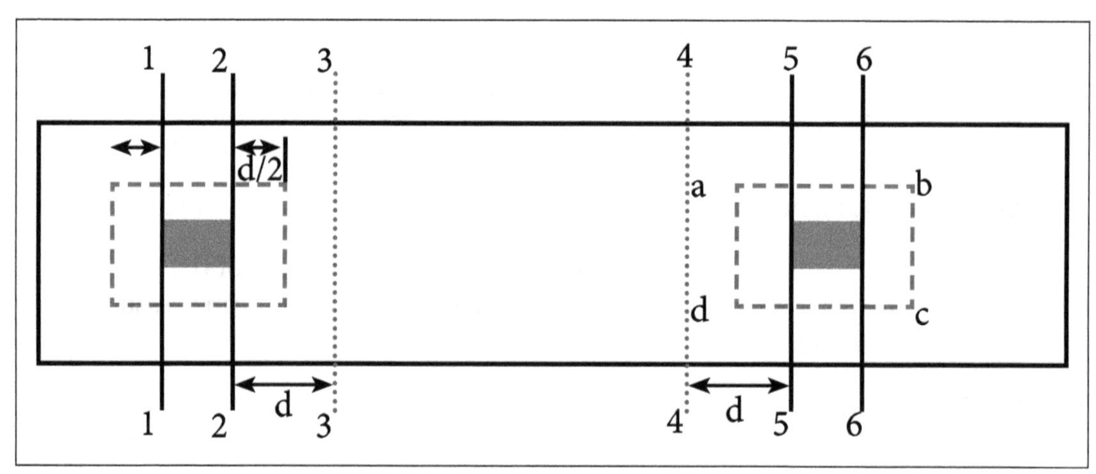

Critical Sections for Moments and Shear.

Section 1-1, 2-2, 5-5 and 6-6 are sections for critical moments.

Section 3-3, 4-4 are sections for critical shear (one way).

Section for critical two way shear is abcd.

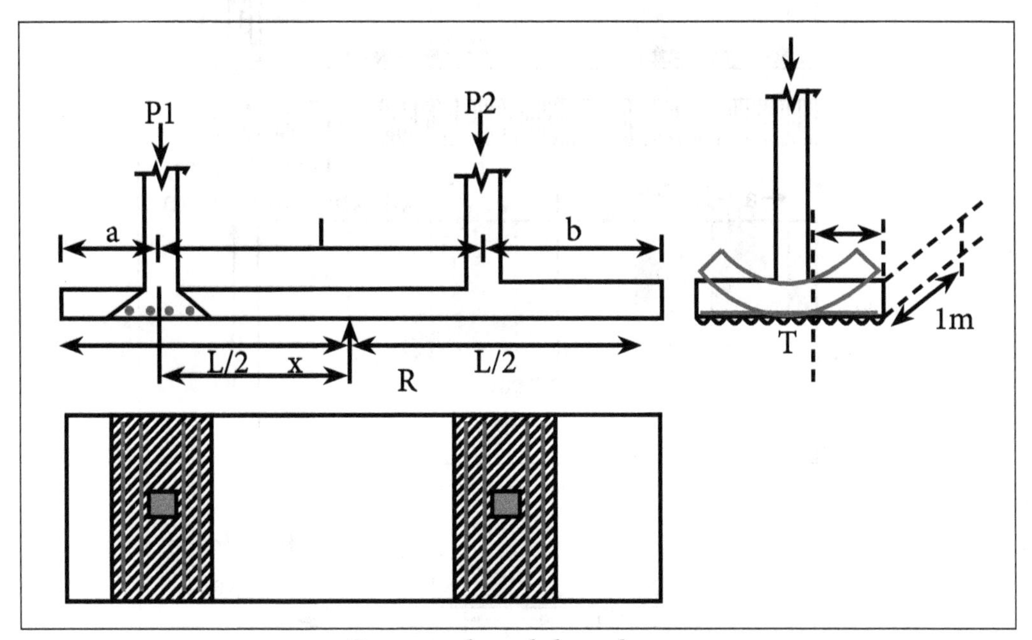

Transverse beam below columns.

The design of rigid rectangular combined footing consists in determining the location of center of gravity of the column loads and using length and width dimensions such that centroid of the footing and the centre of gravity of column loads coincide.

Resulting pressure distribution will be rectangular with the pressure intensity $q = \frac{P_1 + P_2}{B}$ (per unit area).

Column loads may be considered to be concentrated loads and the resulting shear force and bending moment diagrams can be plotted.

Maximum bending moment should be adopted as the design value for the reinforced concrete footing which should also be checked for maximum shear and bond etc.

Beam-Slab Combined Footings

In the case of relatively large footings, providing a uniform large thickness for the entire footing is uneconomical. In such a case, it may be more economical to design a beam-slab footing, in which the footing consists of a base slab stiffened by means of a central longitudinal beam interconnecting the columns as shown in the figure.

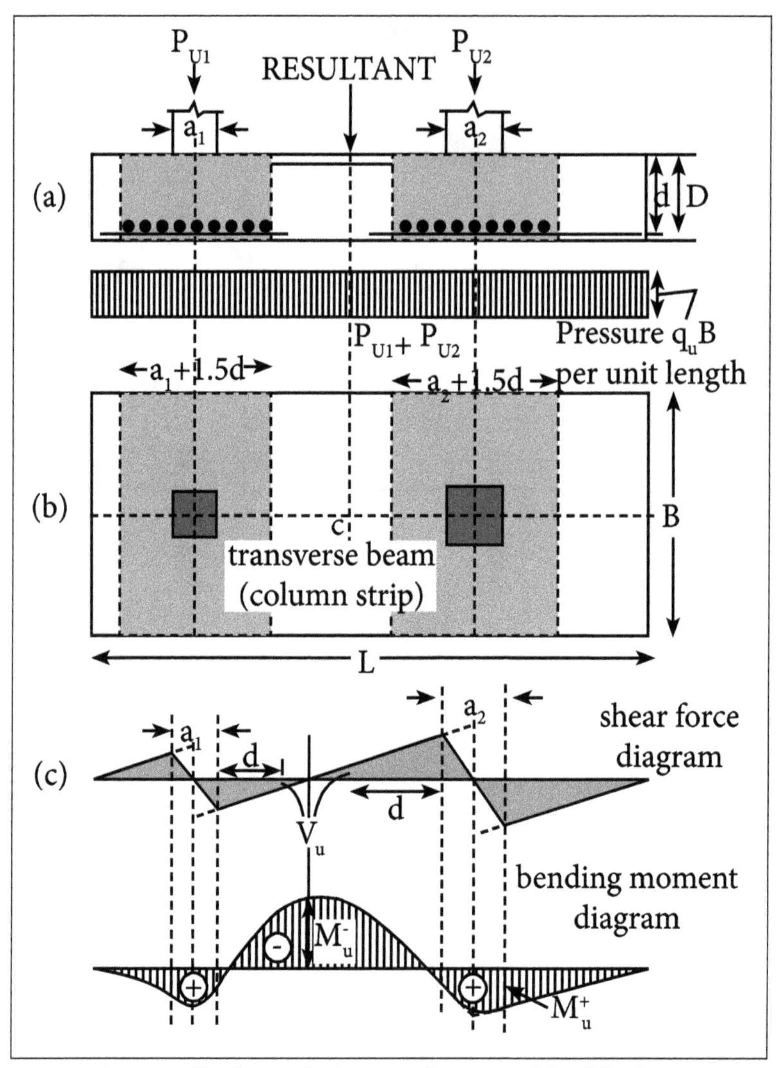

Assumed load transfer in two-column combined footing.

The base slab behaves like a one-way slab, supported by the beam and bends transversely under the uniform soil pressure acting from below. The loads transferred from the slab are resisted by the longitudinal beam.

The size of the beam is generally governed by (one-way) shear at d from the face of the column/pedestal. For effective load transfer, the width of the footing beam should be made equal to the column/pedestal width and it is advantageous to provide a pedestal to the column. The high shear in the beam will usually call for heavy shear reinforcement, usually provided in the form of multi-legged stirrups as shown in the figure below.

The base slab may be tapered (if the span (B-b)/2 is large), for economy. The thickness of the slab should be checked for one-way shear at d (of slab) from the face of the beam.

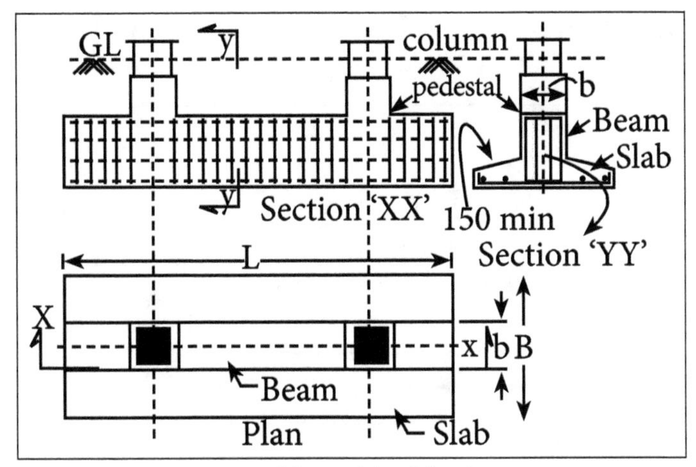

Beam-slab combined footing.

The flexural reinforcement in the slab is designed for the cantilever moment at the face of the beam, and provided at the bottom, as shown in the figure above. Two-way shear is not a design consideration in beam-slab footings. The top and bottom reinforcement in the beam should conform to the longitudinal bending moment diagram and development length requirements should be satisfied.

Design Steps

- Step 1: Initially, locate the point of application of the column loads on the footing.
- Step 2: Then proportion the footing such that the resultant of loads passes through the center of footing.
- Step 3: Compute the area of footing such that the allowable soil pressure is not exceeded.
- Step 4: Then, calculate the shear forces and bending moments at the salient points and hence draw SFD and BMD.
- Step 5: Fix the depth of footing from the maximum bending moment.
- Step 6: Calculate the transverse bending moment and design the transverse section for depth and reinforcement. Then, check for anchorage and shear.
- Step 7: Check the footing for longitudinal shear and design the longitudinal steel.
- Step 8: Design the reinforcement for the longitudinal moment and place them in the appropriate positions.
- Step 9: Check the development length for longitudinal steel. Then, curtail the longitudinal bars for economy.

- Step 10: Finally, draw and detail the reinforcement and prepare the bar bending schedule.

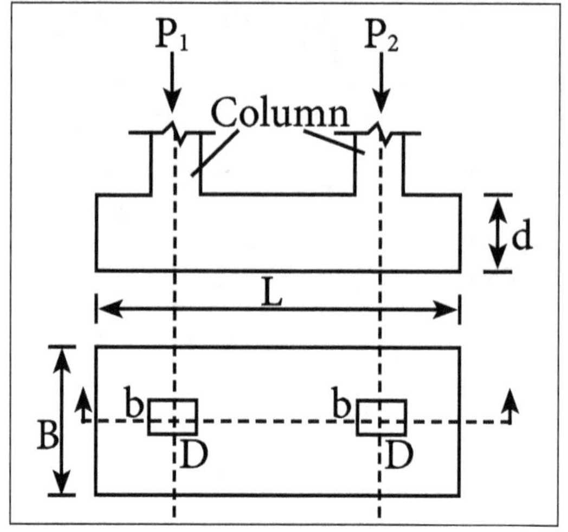

Combined footing without a central beam.

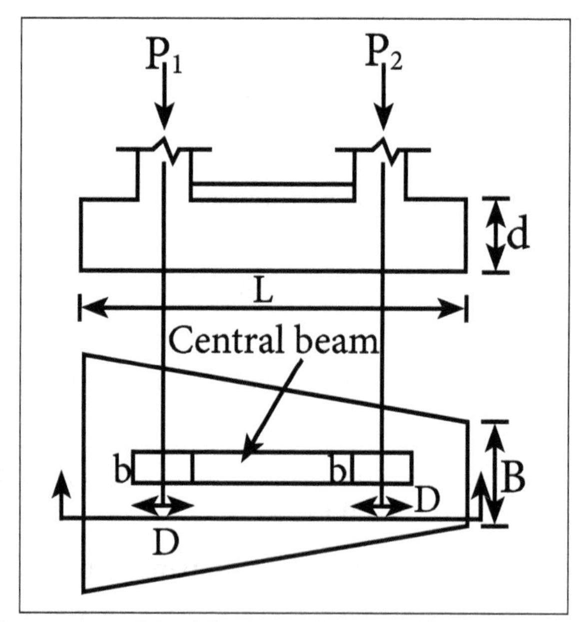

Combined footing with a central beam.

Spacing of the adjacent columns is so close. Hence separate isolated footings are not possible.

Due to inadequate clear space between the two areas of the footings or overlapping areas of the footings, combined footings are the solution for combining two or more columns. Normally, it means a footing combining two columns.

This footings are either rectangular or trapezoidal in plan forms with or without a beam that joins the two columns.

Detailing

Detailing of steel (both longitudinal and transverse) in a combined footing is similar to that of conventional beam-SP-34.

Detailing requirements of beams and slabs should be followed as appropriate-SP-34.

Problem

1. Two interior columns A and B carry 700 kN and 1000 kN loads respectively. Column A is 350 mm × 350 mm and column B is 400 mm × 400 mm in section. The centre to centre spacing between columns is 4.6 m. The soil on which the footing rests is capable of providing resistance of 130 kN/m². Let us design a combined footing by providing a central beam joining the two columns. Use concrete grade M25 and mild steel reinforcement.

And let us also draw to a suitable scale the following:

- The longitudinal sectional elevation.
- Transverse section at the left face of the heavier column.

Plan of the footing:

Solution:

Given:

$$f_{ck} = 25 \text{ N/mm}^2$$

$$f_y = 250 \text{ N/mm}^2$$

$$f_b = 130 \text{ kN/m}^2 \,(\text{SBC})$$

Column A = 350 mm × 350 mm

Column B = 400 mm × 400 mm,

c/c Spacing of columns = 4.6 m,

P_A = 700 kN and P_B = 1000 kN

Required: To design combined footing with central beam joining the two columns.

Ultimate loads,

$$P_{u,A} = 1.5 \times 700 = 1050 \text{ kN},\ P_{u,B} = 1.5 \times 1000 = 1500 \text{ kN}$$

Proportioning of Base Size:

Working load carried by column $A = P_A = 700$ kN

Working load carried by column $B = P_B = 1000$ kN

Self weight of footing 10 % x $(P_A + P_B) = 170$ kN

Total working load = 1870 kN

Required area of footing = A_f = Total load /SBC

$= 1870/130 = 14.38 \text{ m}^2$

Let the width of the footing $= B_f = 2$ m

Required length of footing $= L_f = A_f/B_f = 14.38/2 = 7.19$ m

Provide footing of size 7.2 m X 2m, $A_f = 7.2 \times 2 = 14.4 \text{ m}^2$

For uniform pressure distribution the C.G. of the footing should coincide with the C.G. of column loads. Let x be the distance of C.G. from the centre line of column A.

Then $x = (P_B \times 4.6)/(P_A + P_B) = (1000 \times 4.6)/(1000 + 700)$

= 2.7 m from column A.

If the cantilever projection of footing beyond column A is 'a' then, $a + 2.7 = L_f/2 = 7.2/2$, Therefore a = 0.9 m.

Similarly if the cantilever projection of footing beyond B is 'b' then,

$b + (4.6 - 2.7) = L_f/2 = 3.6$ m,

Therefore b = 3.6 - 1.9 = 1.7 m

The details are shown in the below figure:

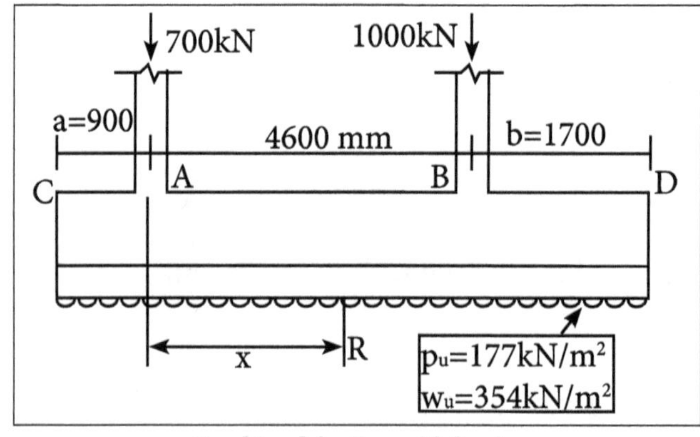

Combined footing with loads.

Total ultimate load from columns $= P_u = 1.5(700 + 1000) = 2550$ kN.

Upward intensity of soil pressure $w_u = P/A_f = 2550/14.4 = 177$ kN/m^2

Design of Slab

Intensity of Upward pressure $= w_u = 177$ kN/m^2

Consider one meter width of the slab (b = 1m)

Load per m run of slab at ultimate = 177 × 1 = 177 kN/m

Cantilever projection of the slab (For smaller column)

=1000 - 350/2 = 825 mm

Maximum ultimate moment $= 177 \times 0.825^2/2 = 60.2$ kN-m.

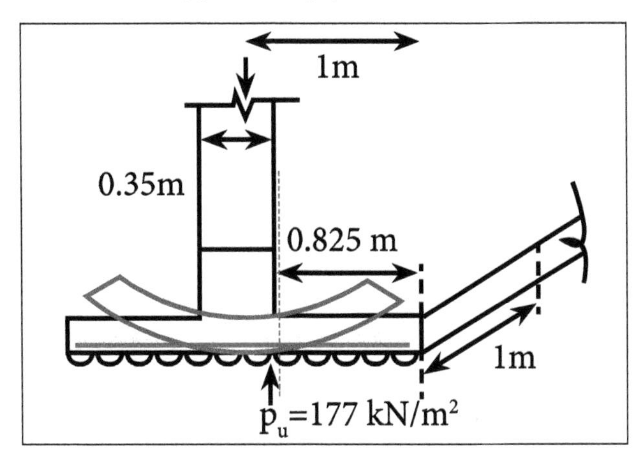

For M25 and Fe 250, $Q_{u,max} = 3.71$ N/mm^2

Required effective depth $= \sqrt{(60.2 \times 10^6/(3.71 \times 1000))} = 128$ mm

Since the slab is in contact with the soil clear cover of 50 mm is assumed.

Using 20 mm diameter bars,

Required total depth = 128 + 20/2 + 50 =188 mm say 200 mm

Provided effective depth = d = (200-50-20)/2 = 140 mm

To find steel,

$M_u/bd^2 = 3.07 < 3.73$, URS

$$M_u = 0.87 f_y A_{st} [d - f_y A_{st}/(f_{ck}, b)]$$

$p_t = 1.7\%$

$A_{st} = 2380 \text{ mm}^2$

Use $\phi 20$ mm diameter bar at spacing = 1000 × 314 / 2380 = 131.93 say 130 mm c/c.

Area provided =1000 × 314 / 130 = 2415 mm²

Check the depth for one - way shear considerations- At 'd' from face.

Design shear force = $V_u = 177 \times (0.825 - 0.140) = 121$ kN

Nominal shear stress = $\tau_v = V_u/bd = 121000/(1000 \times 140) = 0.866$ MPa

Permissible Shear Stress

$P_t = 100 \times 2415 / (1000 \times 140) = 1.7\%$, $\tau_{uc} = 0.772 \text{ N/mm}^2$

Value of k for 200 mm thick slab = 1.2

Permissible shear stress = $1.2 \times 0.772 = 0.926 \text{ N/mm}^2$

$\tau_{uc} > \tau_v$ and hence safe.

The depth may be reduced uniformly to 150 mm at the edges.

Check for Development Length:

$L_{dt} = [0.87 \times 250/(4 \times 1.4)]\ \Phi = 39\ \Phi$

$= 39 \times 20 = 780$ mm

Available length of bar = 825 – 25 = 800mm > 780 mm and hence safe.

Transverse Reinforcement

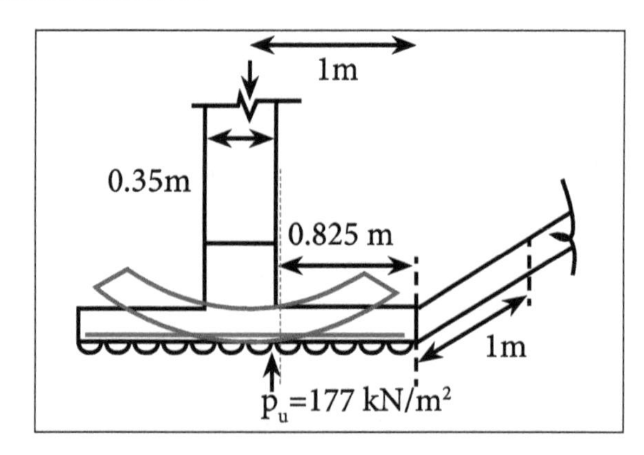

Required $A_{st} = 0.15\ bD/100$

$$= 0.15 \times 1000 \times 200 / 100 = 300\ mm^2$$

Using 8 mm φ bars, Spacing=1000 × 50/300

= 160 mm

Provide distribution steel of φ8 mm at 160 mm c/c, < 300, < 5d.

Design of Longitudinal Beam

Load from the slab will be transferred to the beam.

As the width of the footing is 2 m, the net upward soil pressure per meter length of the beam = w_u = 177 × 2 = 354 kN/m.

Shear Force and Bending Moment

$$V_{AC} = 354 \times 0.9 = 318.6\ kN,\ V_{AB} = 1050 - 318.6 = 731.4\ kN$$

$$V_{BD} = 354 \times 1.7 = 601.8\ kN,\ V_{BA} = 1500 - 601.8 = 898.2\ kN$$

Point of zero shear from left end C.

$$X_1 = 1050 / 354 = 2.97\ m \text{ from C or}$$

$$X_2 = 7.2 - 2.97 = 4.23\ m \text{ from D}$$

Maximum B.M. occurs at a distance of 4.23 m from D,

$$M_{uE} = 354 \times 4.23^2 / 2 - 1500\ (4.23 - 1.7) = -628\ kN.m$$

Bending moment under column $A = M_{uA} = 354 \times 0.9^2/2 = 143.37$ kN.m

Bending moment under column $B = M_{uB} = 354 \times 1.7^2 = 511.5$ kN-m

Let the point of contra flexure be at a distance x from the centre of column A.

Then, $M_x = (1050x) - 354\ (x + 0.9)^2/2 = 0$

Therefore x = 0.206 m and 3.92 m from column A i.e. 0.68 m from B.

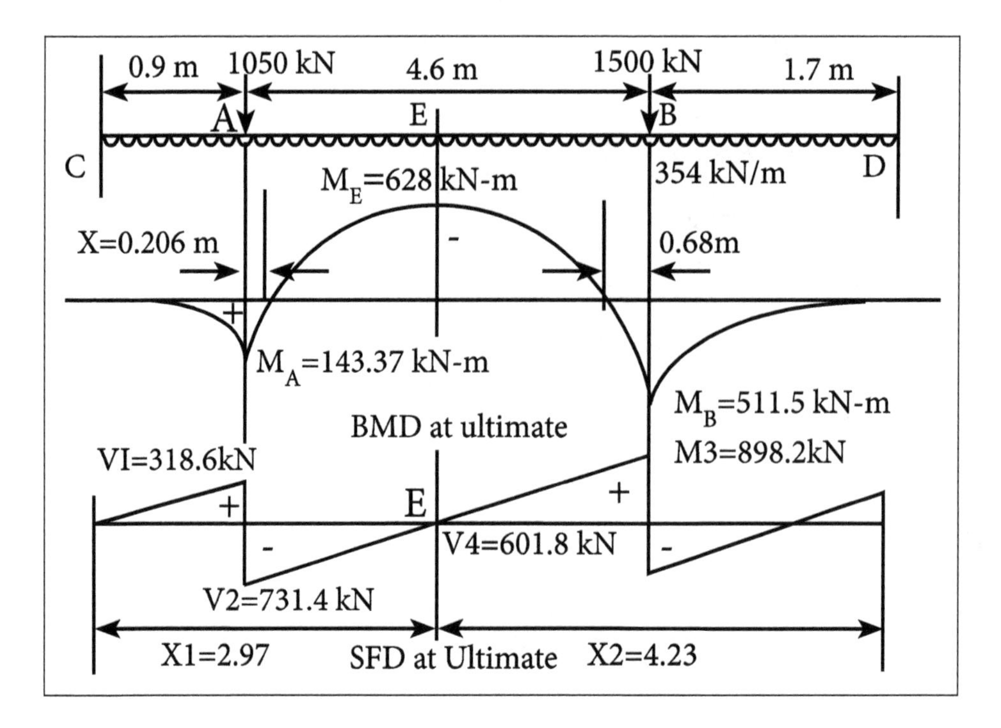

Depth of Beam from B. M.

The width of beam is kept equal to the maximum width of the column i.e. 400 mm. Let us determine the depth of the beam where T- beam action is not available.

The beam acts as a rectangular section in the cantilever portion, where the maximum positive moment = 511.5 kN/m.

$$d = \sqrt{\left(511.5 \times 10^6 / (3.73 \times 400)\right)} = 586 \text{ mm}$$

Provide total depth of 750 mm. Assuming two rows of bars with effective cover of 70 mm.

Effective depth provided = d = 750-70 = 680 mm (Less than 750mm and hence no side face steel is needed).

Check the Depth for Two-way Shear

The heaver column B can punch through the footing only if it shears against the depth of the beam along its two opposite edges and along the depth of the slab on the remaining two edges.

The critical section for two-way shear is taken at distance d/2 (i.e. 680/2 mm) from the face of the column. Therefore, the critical section will be taken at a distance half the effective depth of the slab ($d_s/2$) on the other side as shown in the figure.

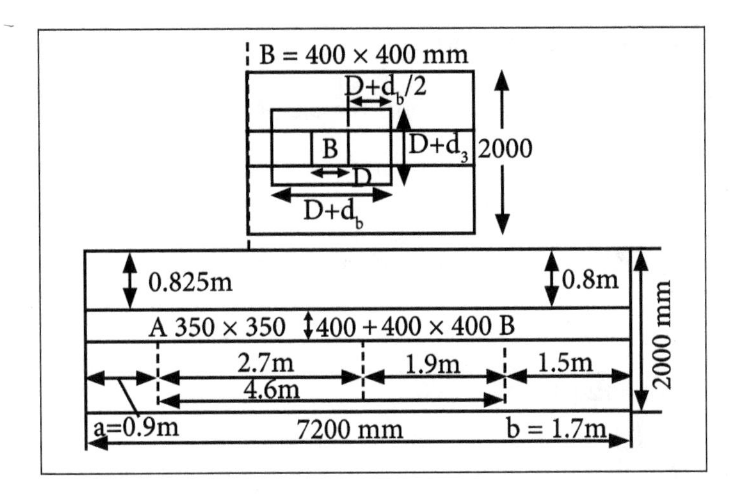

In this case $b = D = 400$ mm, $d_b = 680$ mm, $d_s = 140$ mm

Area resisting two-way shear,

$$= 2(b \times d_b + d_s \times d_s) + 2\,(D + d_b)d_s$$

$$= 2\,(400 \times 680 + 140 \times 140) + 2(400 + 680)\,140 = 885600 \text{ mm}^2$$

Design shear = P_{ud} = Column load – (W_u x area at critical section)

$$= 1500 - 177 \times (b + d_s) \times (D + d_b)$$

$$= 1500 - 177 \times (0.400 + 0.140) \times (0.400 + 0.680) = 1377.65 \text{ kN}$$

$$\tau_v = P_{ud}/b_o d = (1377.65 \times 1000)/885600 = 1.56 \text{ MPa.}$$

Shear stress resisted by concrete = $\tau_{uc} = \tau_{uc} \times K_s$

Where, $\tau_{uc} = 0.25\ \sqrt{f_{ck}} = 0.25\ \sqrt{25} = 1.25 \text{ N/mm}^2$

$$K_s = 0.5 + d/D = 0.5 + 400/400 = 1.5 \leq 1. \text{ Hence } K_s = 1.$$

$$\tau_{uc} = 1 \times 1.25 = 1.25 \text{ N/mm}^2$$

Area of Steel: Cantilever Portion BD

Length of cantilever from the face of column =1.7- 0.4/2 = 1.5 m.

Ultimate moment at the face of column $= 354 \times 1.5^2 / 2 = 398.25$ kN-m.

$$M_{u,max} = 3.71 \times 400 \times 680^2 \times 10^{-6} = 686 \text{ kN-m} > 398.25 \text{ kN-m}$$

Therefore the section is singly reinforced.

$$M_u/bd^2 = 398.25 \times 10^6 / \left(400 \times 680^2\right) = 2.15 < 3.73, \text{ URS}$$

$$P_t = 1.114\%$$

$A_{st} = 3030 \text{ mm}^2$, Provide $3 - \phi 32$ mm + $4 - \phi 16$ mm at bottom face,

Area provided = 3217 mm^2

$$L_{dt} = 39 \times 32 = 1248 \text{ mm.}$$

Curtailment

All bottom bars will be continued up to the end of cantilever. The bottom bars of 3 -φ32 will be curtailed at a distance d (= 680 mm) from the point of contra flexure $(\lambda = 680 \text{ mm})$ in the portion BE with its distance from the centre of support equal to 1360 mm from B.

Cantilever Portion AC

Length of cantilever from the face of column = (900-350)/2 = 725 mm

Ultimate moment = $354 \times 0.725^2 / 2 = 93$ kN-m.

$$M_u / bd^2 = 93 \times 10^6 / \left(400 \times 680^2\right) = 0.52 < 3.73, \text{ URS}$$

$P_t = 0.245\%$ (Greater than minimum steel)

$$A_{st} = 660 \text{ mm}^2$$

Provide $4 - \phi 16$ mm at bottom face, Area provided $= 804 \text{ mm}^2$

Continue all 4 bars of 16 mm diameter throughout at bottom.

Region AB between Points of Contra Flexures

The beam acts as an isolated T- beam:

$b_f = [L_o / (L_o / b + 4)] + b_w$, where,

$$L_o = 4.6 - 0.206 - 0.68 = 3.714 \text{ m} = 3714 \text{ mm}$$

b = actual width of flange = 2000 mm, $b_w = 400$ mm

$$b_f = \left[3714 / \left(3714 / 2000 + 4\right) + 400\right] = 1034 \text{ mm} < 2000 \text{ mm}$$

$D_f = 200$ mm, $M_u = 628$ kN-m

Moment of resistance, M_{uf} of a beam for $x_u = D_f$ is:

$$M_{uf} = \left[0.36 \times 25 \times 1034 \times 200(680 - 0.42 \times 200)\right] \times 10^{-6}$$

$$= 1109 \text{ kN.m} > Mu(= 628 \text{ kN-m})$$

Therefore $X_u < D_f$

$$M_u = 0.87\, f_y\, A_{st}\left(d - f_y\, A_{st} / f_{ck} b_f\right)$$

$$A_{st} = 4542 \text{ mm}^2$$

Provide 5 bars of ϕ 32 mm and 3 bars of ϕ 16 mm,

Area provided $= 4021 + 603 = 4624 \text{ mm}^2 > 4542 \text{ mm}^2$

$$p_t = 100 \times 4624 / (400 \times 680) = 1.7\ \%$$

Curtailment

Consider that $2 - \phi 32$ mm are to be curtailed.

Number of bars to be continued $= 3 - \phi 16 + 3 - \phi 32$

Giving area $= A_{st} = 3016 \text{ mm}^2$

Moment of resistance of continuing bars,

$$M_{ur} = (0.87 \times 250 \times 3016\ (680 - ((250 \times 3016)/(25 \times 400) \times 10^{-6} = 396.6 \text{ kN-m}$$

Let the theoretical point of curtailment be at a distance x from the free end C,

Then, $M_{uc} = M_{ur}$. Therefore, $-354\, x^2 / 2 + 1050\ (x - 0.9) = 396.6$

$x^2 - 5.93x + 7.58 = 0$, Therefore x = 4.06m or 1.86m from C.

Actual point of curtailment = 4.06 + 0.68 = 4.74 m from C or 1.86 - 0.68 = 1.18 m from C.

Terminate 2 - φ 32 mm bars at a distance of 280 mm (= 1180 - 900) from the column A and 760mm (= 5500 - 4740) from column B. Remaining bars 3 - φ32 shall be continued beyond the point of inflection for a distance of 680 mm i.e. 460 mm from column A and up to the outer face of column B. Remaining bars of 3 - φ 16 continued in the cantilever portion.

Design of Shear Reinforcement

Portion between Column i.e. AB:

In this case the crack due to diagonal tension will occur at the point of contra flexure because the distance of the point of contra flexure from the column is less than the effective depth d(= 680mm).

Maximum shear force at $B = V_{u,max} = 898.2$ kN

Shear at the point of contra flexure

$$= V_{uD} - 898.2 - 354 \times 0.68 = 657.48 \text{ kN}$$

$$\tau_v = 657000/(400 \times 680) = 2.42 \text{ MPa} < \tau_{c,max}.$$

Area of steel available $3 - \phi16 + 3 - \phi32$,

$$A_{st} = 3016 \text{ mm}^2$$

$$p_t = 100 \times 3016/(400 \times 680) = 1.1\%$$

$$\tau_c = 0.664 \text{ MPa}$$
$$\tau_v > \tau_c$$

Design shear reinforcement is required.

Using 12 mm diameter 4 - legged stirrups,

Spacing= [0.87 × 250 × (4 × 113)] / (2.42 – 0.664) × 400 = 139 mm say 120 mm c/c.

Zone of shear reinforcements between τ_v to $\tau_c = 0.12$ m from support B towards A.

Maximum shear force at A

$$= V_{u,max} = 731.4 \text{ kN}.$$

Shear at the point contra flexure $= V_{uD} = 731.4 - 0.206 \times 354 = 658.5$ kN

$$\tau_v = 658500/(400 \times 680) = 2.42 \text{ MPa} < \tau_{c,max}$$

Area of steel available = 4624 mm^2, p_t = $100 \times 4624/(400 * 680) = 1.7$ %

$$\tau_{uc} = 0.772 \text{ N/mm}^2,$$

$$\tau_v > \tau_c$$

Design shear reinforcement is required.

Using 12 mm diameter 4 - legged stirrups,

Spacing = 0.87 × 250 × (4 × 113) /(2.42-0.774)×400 =149 mm say 140 mm c/c.

Zone of shear reinforcement, From A to B for a distance as shown in figure.

For the remaining central portion of 1.88 m provide minimum shear reinforcement using 12 mm diameter 2 - legged stirrups at,

Spacing, s = 0.87 × 250 × (2 × 113)/(0.4 × 400) = 307.2 mm, Say 300 mm c/c< 0.75d

Cantilever Portion BD

$$V_{u,max} = 601.8 \text{ kN},$$

$$V_{uD} = 601.8 - 354(0.400/2 + 0.680) = 290.28 \text{ kN}.$$

$$\tau_v = 290280/(400 \times 680) = 1.067 \text{ MPa} < \tau_{c,max}.$$

$$A_{st} = 3217 \text{ mm}^2 \text{ and } p_t = 100 \times 3217/(400 \times 680) = 1.18\%$$

$$\tau_c = 0.683 \text{N/mm}^2$$

$\tau_v > \tau_c$ and $\tau_v - \tau_c < 0.4$. Provide minimum steel.

Using 12 mm diameter 2- legged stirrups,

Spacing = 0.87 × 250 × (2 × 113) /(0.4×400) =307.2 mm, say 300 mm c/c.

Cantilever Portion AC

Minimum shear reinforcement of φ12 mm diameters 2 - legged stirrups at 300mm c/c will be sufficient in the cantilever portions of the beam as the shear is very less.

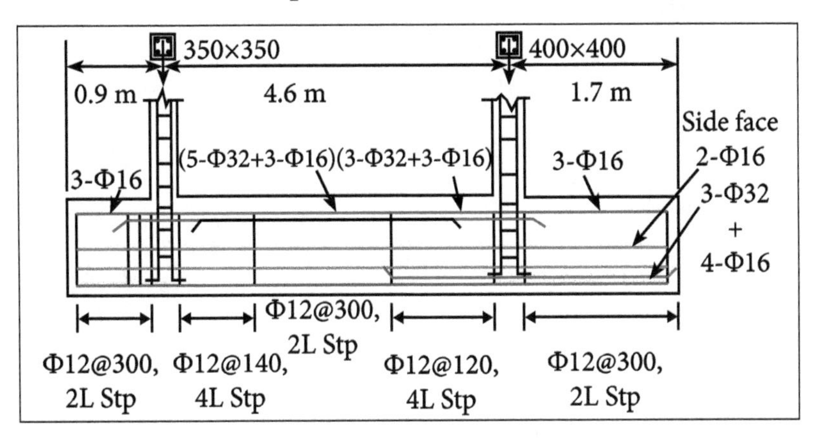

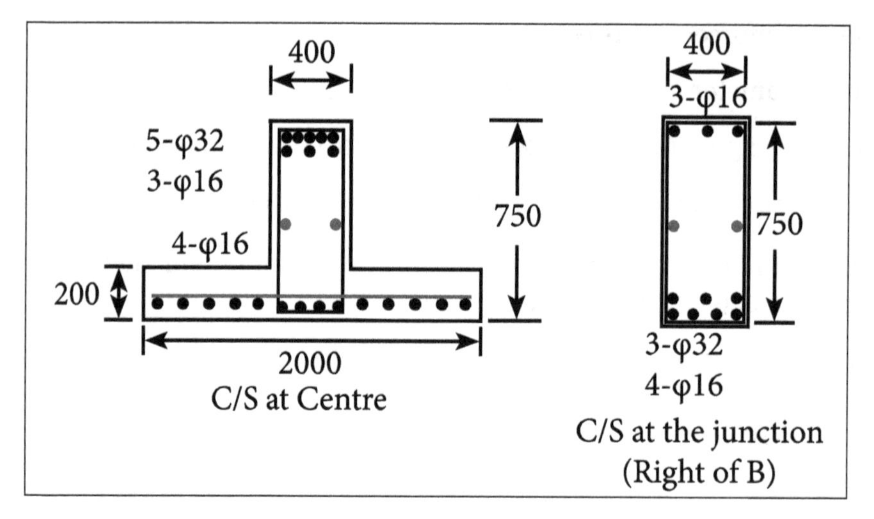

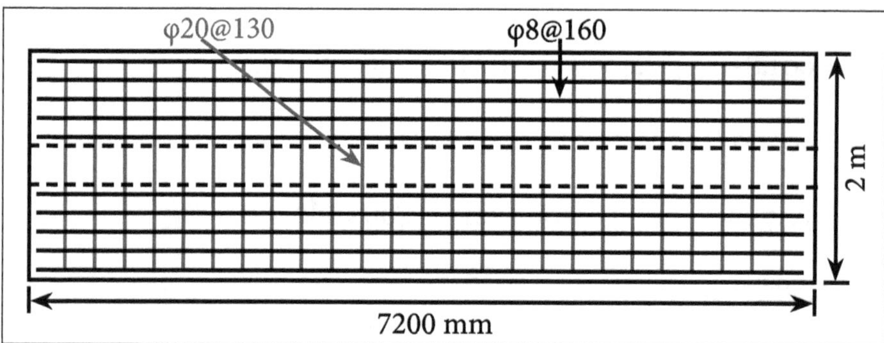

Plan of footing slab.

6

Design and Detailing of Retaining Walls

6.1 Cantilever Type

Retaining walls are generally used to retain earth or such materials to maintain unequal levels on its two faces. The soil on the back face is at a higher level and is called the backfill. Retaining walls are extensively used in the construction of basements below ground level, wing walls of bridge and to retain slopes in hilly terrain roads.

The retaining wall prevents the retained earth to exert a lateral pressure on the wall tending to bend, over-turn and slide the retaining wall. Retaining walls should be designed to resist the lateral earth pressure from the sides and the soil pressure acting vertically on the footing slab integrally built with the vertical slab.

Gravity walls of stone masonry were generally used in the earlier days to retain earthen embankments. The thickness of the masonry walls increased with the height of the earth fill. The advent of reinforced concrete has resulted in thinner retaining walls of different types resulting in considerable reduction of costs coupled with improved aesthetics.

Types of Retaining Wall

Retaining wall can be classified structurally as:

- Cantilever retaining wall.
- Counterfort retaining wall.

Cantilever Retaining Wall

The most common and widely used retaining wall is of the cantilever type comprising the following structural parts as shown in the figure below:

- Vertical stem resisting earth pressure from one side and the slab bends like a cantilever. The thickness of the slab is larger at the bottom and gradually decreases towards the top in proportion to the variation in soil pressure.
- The base slab forming the foundation comprises the heel slab and the toe slab. The heel slab acts as a horizontal cantilever under the combined action of the weight of retained earth from the top and the soil pressure acting from the soffit.

The toe slab also acts as a cantilever under the action of the resulting soil pressure acting upward.

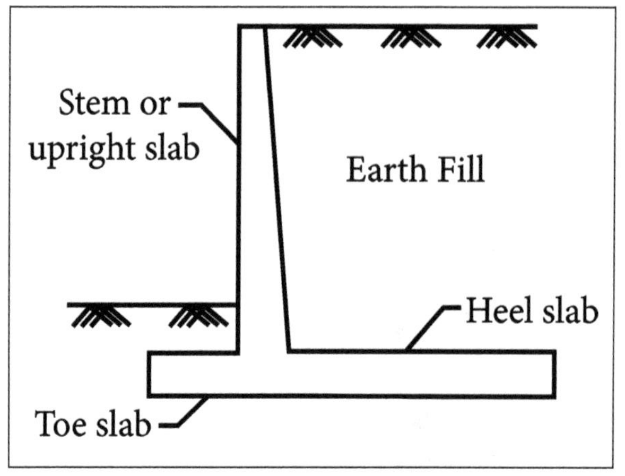

Cantilever Retaining Wall.

The stability of the wall is maintained by the weight of the earth fill on the heel slab together with the self-weight of the structural elements of the retaining wall. Cantilever type retaining walls are adopted for small to medium heights up to 5m.

The stability of retaining walls should be checked against the following conditions:

- The wall should be stable against bearing capacity failure.
- The wall should be stable against overturning.
- The wall should be stable against sliding.

The minimum factor of safety for the stability of a retaining wall:

- The wall should be stable against sliding = 1.5.
- The wall should be stable against Overturning:
    - For Granular Backfill = 1.5.
    - For cohesive backfill = 2.0.
- The wall should be stable against bearing capacity failure.
    - For Granular Backfill = 1.5.
    - For cohesive backfill = 2.0.

Problem

1. Let us determine suitable dimensions of a cantilever retaining wall, which is required

to support a bank of earth 4.0m high above ground level on the toe side of the wall. Consider the backfill surface to be inclined at an angle of 15° with the horizontal. Assume good soil for foundation at a depth of 1.25m below ground level with SBC of 160kN/m². Further assume the backfill to comprise of granular soil with unit weight of 16kN/m³ and an angle of shearing resistance of 30°. Also assume the coefficient of friction between soil and concrete to be 0.5.

Solution:

Given:

$$h = 4.0 + 1.25\text{m}$$

$$\theta = 150$$

$$\hat{o} = 30^\circ$$

$$\gamma_e = 16\text{kN}/\text{m}^3$$

$$q_a = 160\text{kN}/\text{m}^2$$

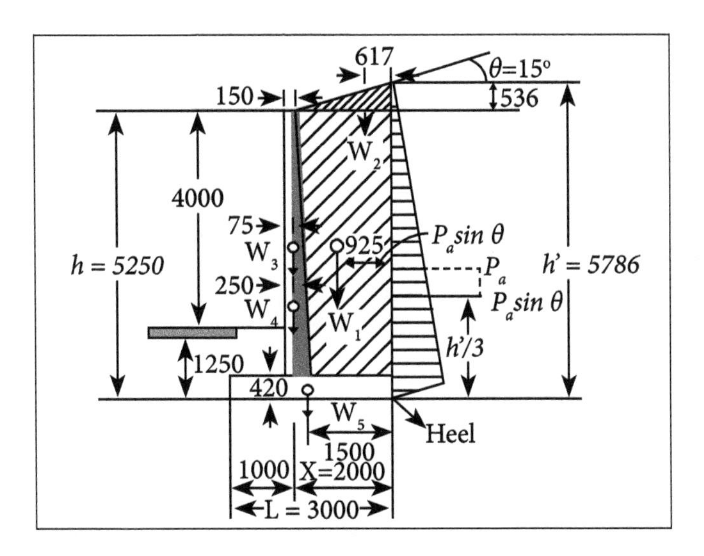

$$d_m = \frac{q_a}{\gamma_e}\left(\frac{1-\sin\phi}{1+\sin\phi}\right)^2 = \frac{160}{16}\left(\frac{1}{3}\right)^2 = 1.11\text{ m}$$

$$\mu = 0.5$$

Earth pressure coefficient,

$$C_a = \left[\frac{\text{Cos}\theta - \sqrt{\text{Cos}^2\theta - \text{Cos}^2\phi}}{\text{Cos}\theta + \sqrt{\text{Cos}^2\theta - \text{Cos}^2\phi}}\right]\text{Cos}\theta = 0.373$$

$$C_p = \frac{1+\text{Sin}\phi}{1-\text{Sin}\phi} = 3.0$$

Preliminary Proportioning

Thickness of footing base slab = 0.08h = 0.08 × 5.25 = 0.42m

Provide a base thickness of 420mm for base slab.

Assume stem thickness of 450mm at the base of stem tapering to 150mm at the top of the wall.

For economical proportioning of length 'L', assume vertical reaction R at the footing base to be in line with front face of the stem.

$$X = \left(\sqrt{\frac{C_a}{3}}\right)h' = \sqrt{\frac{0.373}{3}}(5.25+0.4)$$

= 2.0m [Where 0.4m is assumed as height above wall].

Assuming a triangular base pressure distribution,

$$L = 1.5 \times 2 = 3.0 \text{ m}$$

For the assumed proportions, the retaining wall is checked for stability against overturning and sliding.

Stability Against Overturning

Force ID	Force (kN)	Distance from heel (m)	Moment (kNm)
W_1	16(1.85)(5.25 - 0.42) = 143.0	0.925	132.3
W_2	16(1.85)(0.5 × 0.536) = 7.9 [2tan15° = 0.536]	0.617	4.9
W_3	25(0.15)(5.25-0.42) = 18.1	1.925	34.8
W_4	(25-16)(4.83)(0.5 × 0.3) = 6.5	1.75	11.4
W_5	25(3)(0.42) = 31.5	1.50	47.2
P_a Sinθ	25.9	0	0
Total	W = 232.9		M_w = 230.6

- P_a = Active pressure exerted by retained earth on wall [both wall and earth move in same direction].

- P_p = Passive pressure exerted by wall on retained earth [both move in opposite direction].
- C_a = Same for dry and submerged condition, since ϕ for granular soil does not change significantly.

Force due to active pressure,

$$P_a = C_a \cdot \gamma_e . h'^2/2$$

where,

$$h' = h + X \tan\theta$$

$$= 5250 + 2000 \tan 15° = 5786 \text{ mm}$$

$$Pa = 0.373(16)(5.786)^2/2 = 99.9 \text{ kN [per m length of wall].}$$

FOS = 0.9 × Stabilizing force or moment/Destabilizing force or moment

$$\text{Therefore, FOS(overturning)} = \frac{0.9 \text{ Mr}}{\text{Mo}} \geq 1.4$$

$$\text{Overturning moment, } M_o = (P_a \cos\theta)h'/3 = 96.5(5.786/3) = 186.1 \text{ kNm}$$

Distance of resultant vertical force from heel,

$$X_w = M_w/W = 230.6/232.9 = 0.99 \text{ m}$$

Stabilizing moment (about toe),

$$M_r = W(L - X_w) + P_a \text{Sin}\theta(L) = 232.9(3 - 0.99) + 77.6 = 468.1 \text{ kNm[per m length of wall]}$$

$$FOS_{(overturning)} = \frac{0.9 \text{Mr}}{\text{Mo}} = \frac{0.9 \times (468.1 + 77.6)}{186.1} = 2.26 > 1.40$$

Soil Pressure at Footing Base

Resultant vertical reaction, R = W = 232.9kN [per m length of wall].

Distance of R from heel, $L_R = (M_w + M_o)/R = (230.6 + 186.1)/232.9 = 1.789\text{m}$

Eccentricity, $e = L_R - L/2 = 1.789 - 3/2 = 0.289\text{m} < L/6 -> [0.5 \text{ m}]$

Hence, the resultant lies within the middle third of the base, which is desirable.

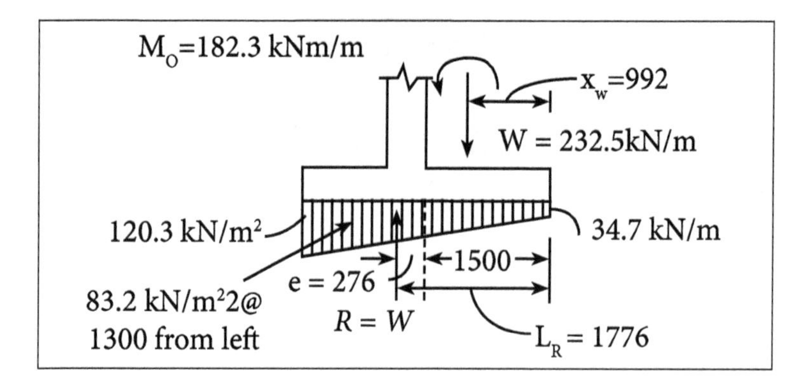

Calculation of Soil pressure.Maximum pressure as base,

$$q_{max} = \frac{R}{L}\left(1+\frac{6e}{L}\right) = \frac{232.9}{3}(1+0.578) = 122.5 \text{ kN/m}^2 < q_q$$

$\left[\text{where } q_a = 160 \text{ kN/m}^2\right]$ [Safe].

$$q_{min} = \frac{R}{L}\left(1-\frac{6e}{L}\right) = \frac{232.9}{3}(1-0.578) = 32.8 \text{ kN/m}^2 < 0$$

[No tension develops] [Safe].

Stability Against Sliding

Sliding force, $P_a \cos\theta = 96.5$ kN

Resisting force, F = μR = 0.5 × 232.9 = 116.4kN [Ignoring passive pressure on toe side].

$$(F.S)_{sliding} = \frac{0.9F}{P_a} = \frac{0.9 \times 257.9}{216.0} = 1.075 < 1.4$$

Hence, a shear key may be provided.

Assume a shear key 300mm × 300mm at a distance of 1300mm from toe as shown in the figure.

$$h_2 = 950 + 300 + 1300 \tan 30° = 2001 \text{ mm}$$

$$Pp = C_p \cdot \gamma_e \cdot \left(h_2^{\,2} - h_1^{\,2}\right)/2 = 3 \times 16 \times \left(2.001^2 - 0.950^2\right)/2 = 74.11 \text{ kN.}$$

$$FOS_{(Sliding)} = \frac{0.9(116.4 + 74.44)}{96.5} = 1.78 > 1.4 \quad [\text{SAFE}]$$

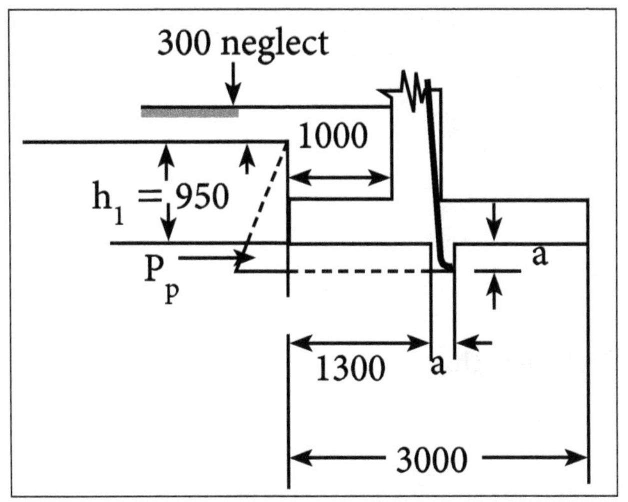

Design of shear key.

Design of Toe Slab

Assuming a clear cover of 75mm and 16mm φ used,

$$d = 420 - 75 - 8 = 337\text{mm}$$

$$V_u = 1.5\left[\frac{112+81.9}{2}\right]\times(1-0.337) = 96.42 \text{ kN}$$

$$M_u = 1.5\{(81.9 \times 12/2)+(112-81.9)\times1/2\times12\times2/3\}$$

$$= 76.48 \text{ kNm/m length}$$

Nominal shear stress,

$$\tau_v = \frac{V_u}{bd} = \frac{96.42\times10^3}{1000\times337} = 0.286 \text{ N/mm}^2$$

Using M_{20} concrete,

For $\tau_c = 0.29 \text{ N/mm}^2$, $p_t(\text{required}) = 0.2\%$

For p_t = 0.2%, $A_{st} = 0.2/100 \times 1000 \times 337 = 674 \text{ mm}^2/\text{m}$

Spacing $= \dfrac{1000\times\pi\times16^2/4}{674} = 298$ mm

Provide 16mm φ @ 290mm c/c at the bottom of toe slab.

$$L_d = \frac{\phi\sigma_s}{4\tau_{bd}} = \frac{(16).0.87\, f_y}{4\tau_{bd}} = 16\times47 = 752 \text{ mm, beyond face of stem.}$$

Since length available is 1m, no curtailment is sorted.

Design of Heel Slab

$$V_u = 1.5\left[\frac{82.54+128.6}{2}\right]\times(1.55-0.337) = 128.06 \text{ kN}$$

$$M_u = 1.5\{(82.54\times1.55^2/2)+(128.6-82.54)\times1/2\times1.55^2\times2/3\}$$

= 203.96 kNm/m length

Nominal shear stress,

$$\tau_v = \frac{V_u}{bd} = \frac{128.06\times10^3}{1000\times337} = 0.38 \text{ N/mm}^2$$

Using M20 concrete,

For a $\zeta_c = 0.39 \text{ N/mm}^2$, $p_t(\text{required}) = 0.3\%$

$$K = \frac{M_u}{bd^2} = \frac{203.96\times10^6}{1000\times337^2} = 1.8 \text{ N/mm}^2$$

For $p_t = 0.565\%$, $A_{st} = 0.565/100 \times 1000 \times 337 = 1904.05 \text{ mm}^2/\text{m}$

$$\text{Spacing} = \frac{1000\times\pi\times16^2/4}{1904.05} = 105.61 \text{ mm}$$

Provide 16mm φ @ 100mm c/c at bottom of toe slab.

$$L_d = \frac{\phi\sigma_s}{4\tau_{bd}} = \frac{(16).0.87\,f_y}{4\tau_{bd}} = 16\times47 = 752 \text{ mm, beyond face to stem.}$$

Since length available is 1.55m, no curtailment is sorted.

Design of Vertical Stem

Height of cantilever above base = 5250 – 420 = 4830mm

Assume a clear cover of 50mm and 16mm φ bar,

d (at base)= 450 – 50 – 16/2 = 392mm

$$M_u = 1.5(C_a.\gamma_e.h^3/6) = 1.5(1/3)(16\times4.92^3/6) = 150.24 \text{ kNm.}$$

$$K = \frac{Mu}{bd^2} = \frac{150.24 \times 10^6}{1000 \times 4.92^2} = 1\,N/mm^2.$$

$$p_t = 0.3\%,\ A_{st} = 0.295/100 \times 1000 \times 392 = 1200\ mm^2.$$

Spacing = 1000 x 201/1200 = 160mm

Provide 16mm φ @ 160mm c/c in the stem, extending into the shear key upto 47φ = 752mm.

Check for shear: [at 'd' from base]

$$V_{u\,(stem)} = 1.5\left(C_a.\gamma_e.Z^2/2\right) = 1.5\left(1/3 \times 16 \times (4.83 - 0.392)^2/2\right) = 53.83\ kN$$

$$\tau_v = \frac{V_u}{bd} = \frac{53.83 \times 10^3}{1000 \times 392} = 0.135\ N/mm^2 < \tau_c$$

Where $\zeta_c = 0.39\ N/mm^2$ for $p_t = 0.3\%$

Hence, safe.

Curtailment of Bars

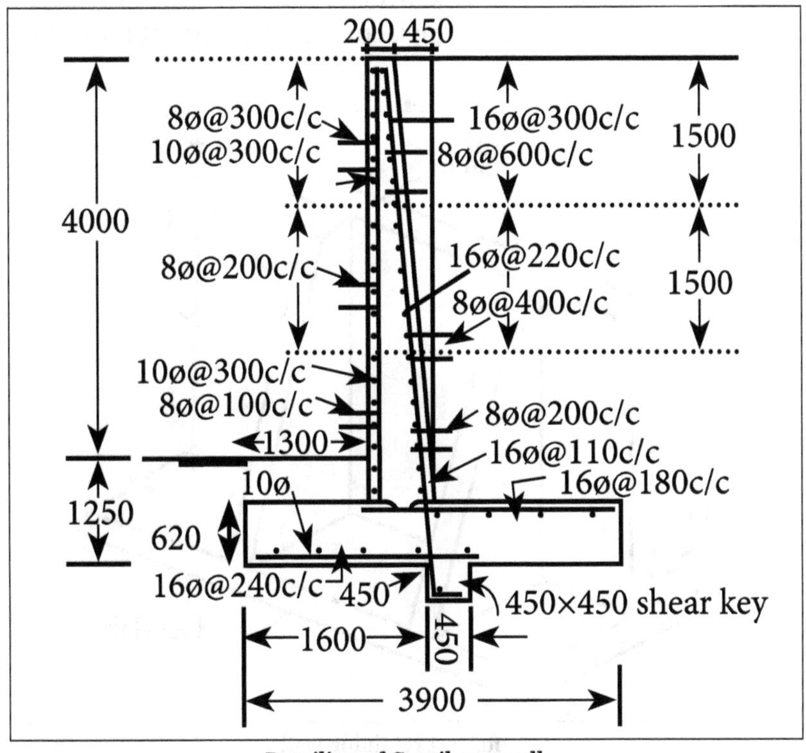

Detailing of Cantilever wall.

Curtailments of bars in stem are done in two stages. At $1/3^{rd}$ and $2/3^{rd}$ height of the stem above base.

Temperature and Shrinkage Reinforcement:

$$A_{st} = 0.12/100 \times 10 \times 450 = 540 \text{ mm}^2$$

For first $1/3^{rd}$ height, provide 2/3rd of bar near front face (exposed to weather) and $1/3^{rd}$ near rear face.

For second $1/3^{rd}$ height, provide $1/2^{nd}$ the above.

For third $1/3^{rd}$ height, provide $1/3^{rd}$ of I case.

Provide nominal bars of 10mm @ 300mm c/c vertically near front face.

6.1.1 Counterfort Type

Counterfort retaining walls are used for large heights exceeding five meters of earth fill. In counterfort retaining wall, the vertical stem is designed as a continuous slab spanning between the counterforts. The Counter-forts are designed as cantilever beams from the base slab.

When the height of earth to be retaining exceeds 5m, the bending moment developed in the stem, heel and toe slabs are very large which results in large thickness of structural elements and becomes uneconomical. Thus, counterfort type retaining wall is adopted for larger heights.

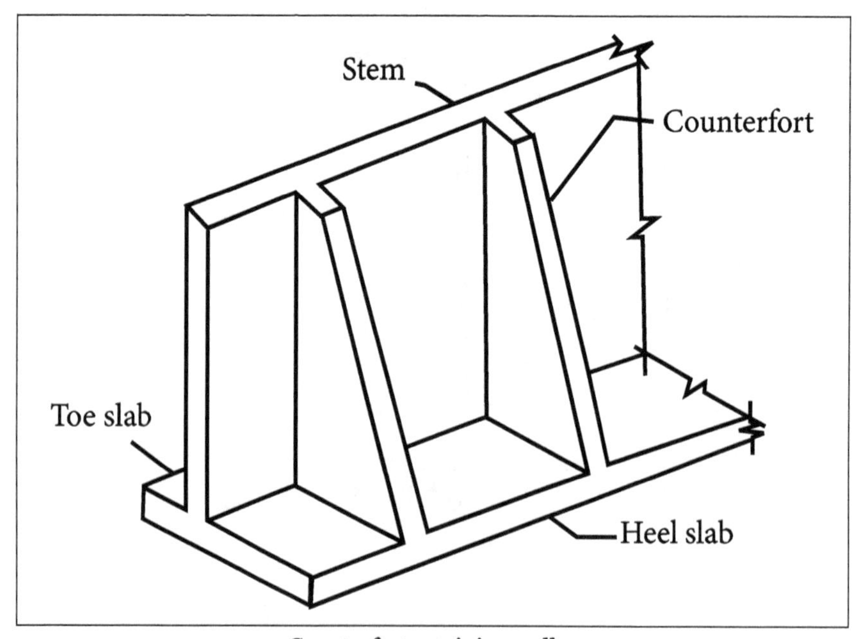

Counterfort retaining walls.

Preliminary proportioning of counterfort retaining wall:

- Thickness of heel slab and stem = 5% of Height of wall.
- Thickness of toe slab [buttress not provided] = 8% of Height of wall.
- Thickness of counterfort = 6% of height of wall.
- Thickness of any component be less than 300mm.
- Spacing of counterfort = 1/3rd to 1/2nd of Height of wall.
- Each panel of stem and heel slab are designed as two way slab with one edge free (one way continuous slab).
- The toe slab is designed as:
    - Cantilever slab when buttress is not provided.
    - One way continuous slab, when buttress is provided.

Counterfort is a triangular shaped structure which is designed similar to a T-Beam as vertical cantilever with varying depth (stem acts as flange). The main reinforcement is along the sloping side. Stirrups are provided in the counterfort to secure them firmly with the stem. Additional ties are provided to securely tie the counterfort to the heel slab.

Problem

1. Let us design a suitable counterfort retaining wall to support a leveled backfill of height 7.5m above ground level on the toe side. Assume good soil for the foundation at a depth of 1.5m below ground level. The SBC of soil is $170 kN/m^2$ with unit weight as $16kN/m^3$. The angle of internal friction is $\varphi = 30^\circ$. The coefficient of friction between the soil and concrete is 0.5. Use M25 concrete and Fe415 steel.

Solution:

Given:

Leveled backfill of height - 7.5m above ground level on the toe side.

Depth - 1.5m below ground level.

SBC of soil – $170\ kN/m^2$

Unit weight – $16kN/m^3$

$\varphi - 30^\circ$

Coefficient of friction between the soil and concrete - 0.5.

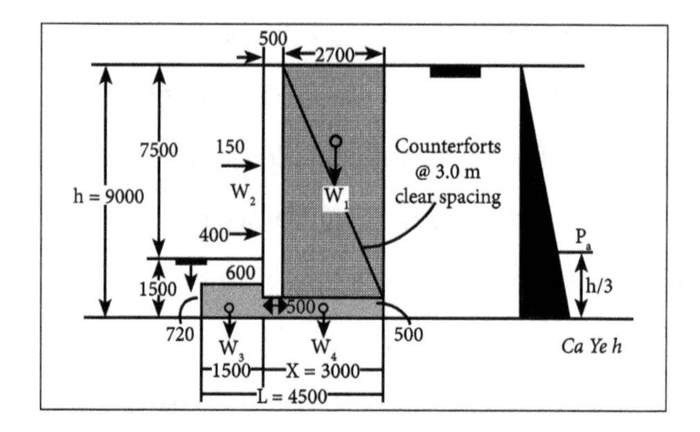

Minimum depth of foundation =

$$\frac{P}{\gamma}\left(\frac{1-\sin\varphi}{1+\sin\varphi}\right)^2 = \frac{170}{16}\left(\frac{1-\sin 30}{1+\sin 30}\right)^2 = 1.181\text{ m} < 1.5\text{ m}$$

Depth of foundation = 1.5m

Height of wall = 7.5 + 1.5 = 9m

Thickness of heel and stem = 5% of 9m = 0.45m ≈ 0.5m

Thickness of toe slab = 8% of 9m = 0.72m

$$X_{min} = \left(\sqrt{\frac{C_a}{3}}\right)h' = \sqrt{\frac{0.333}{3}}(9) = 3.0\text{ m} \quad \left[\text{As } C_a = 1/3 \ \& \ C_p = 3\right]$$

$$L_{min} = 1.5 \times 3 = 4.5\text{ m}$$

Thickness of counterfort = 6% of 9 = 0.54m

Stability Conditions

Earth Pressure Calculations

Force ID	Force (kN)	Distance from heel (m)	Moment (kNm)
W_1	16(7.5+1.5-0.5)(2.5) = 340	(3 - 0.5)/2 = 1.25	425
W_2	25(0.5)(9 – 0.5) = 106.25	0.5/2 + 2.5 = 2.75	292.18
W_3	25(0.5)(3) = 37.5	1.5	56.25
W_4	25(1.5)(0.72) = 27	1.5/2 + 3 = 3.75	101.25
Total	W = 510.75		M_w = 874.69

$$X_w = 874.69/510.75 = 1.713 \text{ m}$$

$$FOS_{(overturning)} = 0.9\ M_r/M_o$$

Where,

$$M_o = Pa.h/3 = C_a.\gamma_e.h^3/6 = 0.33 \times 16 \times 9^3/6 = 647.35 \text{ kNm.}$$

$$M_r = (L - X_w).W = 510.75(4.5 - 1.713) = 1423.6 \text{ kNm.}$$

$$FOS_{(overturning)} = 1.98 > 1.4$$

Hence, section is safe against overturning.

Sliding:

$$FOS_{(sliding)} = 0.9(\mu R)/P_a \cos\theta$$

$$F = \mu R = 0.5 \times 510.75 = 255.375 \text{ kN}$$

$$P_a = C_a.\gamma_e.h^2/2 = 215.784$$

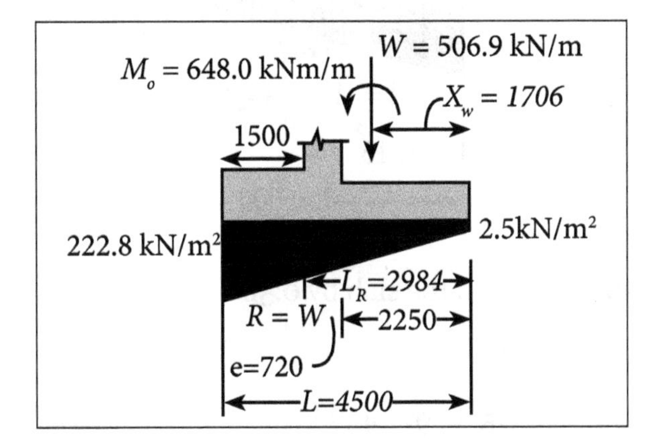

Base Pressure Calculation

$$q_{max} = \frac{R}{L}\left(1 + \frac{6e}{L}\right) = \frac{510.75}{4.5}\left(1 + \frac{6 \times 0.73}{4.5}\right) = 223.97 \text{ kN/m}^2 > q_q$$

Where $q_a = 170$ KN/m², Unsafe.

$$q_{min} = \frac{R}{L}\left(1 - \frac{6e}{L}\right) = \frac{510.75}{4.5}\left(1 - \frac{6 \times 0.73}{4.5}\right) = 3.027 \text{ kN/m}^2 > 0$$

No tension develops. Hence, safe.

Where,

$$L_R = (M + M_o)/R$$
$$e = L_R - L/2$$
$$L_R = (874.688 + 647.352)/510.75 = 2.98 \text{ m}$$
$$e = L_R - L/2 = 2.98 - (4.5/2) = 0.73 < L/6\ (0.75\text{m})$$

Since the maximum earth pressure is greater than SBC of soil, the length of base slab has to be increased preferably along the toe side. Increase the toe slab by 0.5m in length.

$$\sum_W = 510.75 + (0.5 \times 25 \times 0.72) = 519.75 \text{ kN}$$

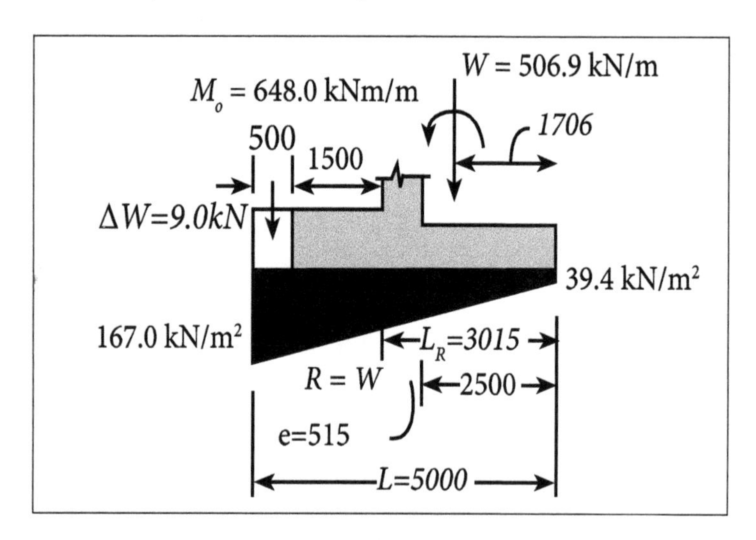

Additional load due to increase in toe slab by 0.5m is,

Moment = 0.5/2 + 4.5 = 4.75 m

$$\sum M = 874.69 + 42.75 = 917.438 \text{ kNm.}$$

$$L_R = (M_o + M)/R = (917.438 + 647.352)/519.75 = 3.011 \text{ m}$$
$$e = L_R - L/2 = 3.011 - (5/2) = 0.511 \text{ m} < L/6\ (0.833 \text{ m})$$

$$q_{max} = \frac{R}{L}\left(1 + \frac{6e}{L}\right) = \frac{519.75}{5}\left(1 + \frac{6 \times 0.511}{5}\right) = 166.32 \text{ kN/m}^2 < q_q$$

where,

$$q_a = 170 \text{ KN/m}^2\text{, Unsafe.}$$

$$q_{min} = \frac{R}{L}\left(1 - \frac{6e}{L}\right) = \frac{519.75}{5}\left(1 - \frac{6 \times 0.511}{5}\right) = 41.58 \text{ kN/m}^2 > 0$$

No tension develops. Hence, safe.

$$FOS_{(Sliding)} = 0.9(\mu R)/P_a = 0.9(0.5 \times 519.75)/215.784 = 1.08 < 1.4.$$

Hence, the section is not safe against sliding. Shear key is provided to resist sliding.

Assume shear key of size 300 × 300 mm.

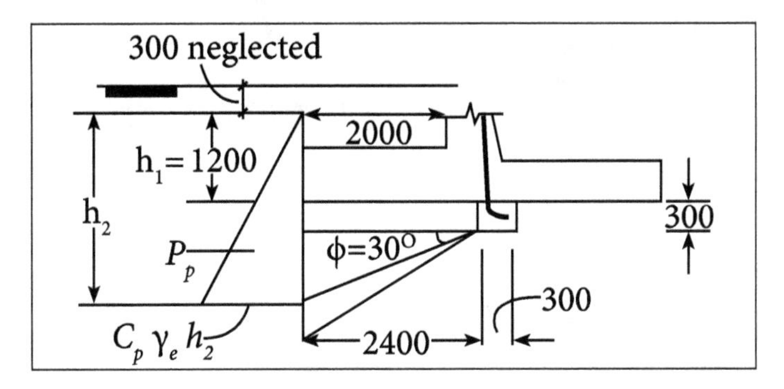

$$P_{ps} = C_p.\gamma_e.(h_2^2 - h_1^2)/2 = 164.5 \text{ kN/m}$$

$$FOS_{(sliding)} = 0.9(\mu R + P_{ps})/Pa = 1.77 > 1.4$$

$$[\text{where, } h_1 = 1.2\text{m, } h_2 = 1.2 + 0.3 + 1.39 = 2.88 \text{ m}]$$

Hence, section is safe in sliding with shear key 300 × 300 mm.

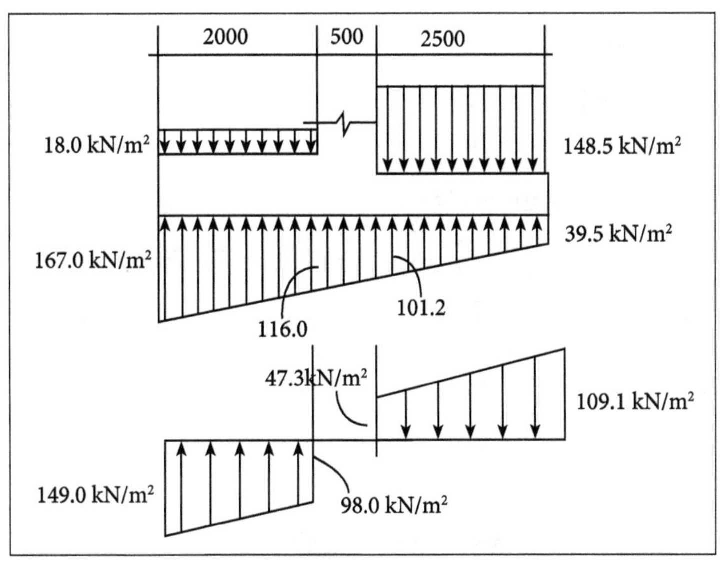

Net soil pressure on base slab.

Design of Toe Slab

Effective cover = 75 + 20/2 = 85 mm.

Toe slab is designed similar to cantilever slab with maximum moment at front face of the stem and maximum shear at 'd' from the front face of the stem.

$$d = 720 - 85 = 635 \text{ mm.}$$

$$M = \left(80.38 \times 2^2/2\right) + \left(½ \times 2 \times 49.94 \times 2/3 \times 2\right) = 160.76 + 122.76 = 227.35 \text{ kNm.}$$

$$\text{SF at } 0.635\text{m} = 49.94/2 \times 0.635 = 15.606 \text{ kN}$$

Area of trapezoid $= ½.h.(a+b) = ½(2-0.635)(130.32+95.98) = 154.44$ kN

Factored SF = 231.66kN

Factored Moment = 341.02 kNm.

$$K = M_u/bd^2 \rightarrow A_{st} = 1551.25 \text{ mm}^2 \rightarrow \text{Spacing}$$
$$= 1000a_{st}/A_{st} \rightarrow 16 \text{ mm } @125 \text{ mm c/c.}$$

Transverse Reinforcement

$$= 0.12\% \text{ of c/s}$$
$$= 0.12/100 \times 1000 \times 720 = 864 \text{ mm}^2$$

Provide 10mm @100mm c/c.

Design of Heel Slab

The heel slab is designed as an one way continuous slab with moment $wl^2/12$ at the support and $wl^2/16$ at the mid span.

The maximum shear at the support is w(l/2 – d).

The maximum pressure at the heel slab is considered for the design.

Moment at the support, $M_{sup} = wl^2/12 = 106.92 \times 2.5^2/12 = 55.688$ kNm.

Moment at the midspan, $M_{mid} = wl^2/16 = 41.76$ kNm

The maximum pressure acting on the heel slab is taken as 'w' for which the A_{st} required at midspan and support are found.

Factored $M_{sup} = 83.53 \text{ kNm} \rightarrow A_{st} = 570.7 \text{ mm}^2$

Factored M_{mid} = 62.64 kNm $\rightarrow$ Ast = 425.4 mm^2

Using 16mm φ bar, Spacing $= 1000_{ast} / A_{st} = 110.02$ mm $\rightarrow$ Provide 16mm @ 110 mm c/c.

At midspan, spacing = 156.72 mm $\rightarrow$ Provide 16mm @ 150 mm c/c.

Transverse reinforcement = 0.12% of c/s = $0.12/100 \times 1000 \times 500 = 600$ mm^2

For 8mm bar, Spacing = 83.775 mm $\rightarrow$ Provide 8 mm @ 80 mm c/c.

Check for Shear:

Maximum shear = w (l/2 – d) = 107 (2.5/2 – 0.415) = 89.345 kN

Factored shear force = 134.0175 kN

$$\zeta_v = 0.33 \text{ N/mm}^2$$
$$\zeta_c = 0.29 \text{ N/mm}^2$$
$$\zeta_{cmax} = 3.1 \text{ N/mm}^2$$

∴ Depth has to be increased.

Design of Stem:

The stem is also designed as one way continuous slab with support moment, $wl^2/12$ and midspan moment, $wl^2/16$. For the negative moment at the support, reinforcement is provided at the rear side and for positive moment at midspan, reinforcement is provided at the front face of the stem.

The maximum moment varies from a base intensity of

$$K_a . \gamma_e . h = 1/3 \times 16 \times (9 - 0.5) = 45.33 \text{ kN/m}$$

$$M_{sup} = wl^2/12 = 1.5 \times 45.33 \times 3.54^2/12 = 71 \text{ kNm}$$

$$M_{mid} = wl^2/16 = 1.5 \times 45.33 \times 3.54^2/16 = 53.26 \text{ kNm}$$

Effective depth = 500 - (50 + 20/2) = 440mm

A_{st} at support = 1058 mm^2

For 16 mm φ, Spacing = 190 mm.

Provide 16 mm @ 190 mm c/c.

A_{st} at midspan =718 mm^2

For 16 mm φ, Spacing = 280 mm.

Provide 16 mm@ 280 mm c/c.

Maximum SF = w (l/2 –d) = 60.29kN

Factored SF = 90.44kN

Transverse reinforcement = 0.12% of c/s → 8 mm @ 80 mm c/c,

$$\zeta_v = 0.188 \text{ N/mm}^2$$
$$\zeta_c = 0.65 \text{ N/mm}^2$$
$$Z_{c\ max} = 3.1 \text{ N/mm}^2$$

∴ Safe in Shear.

Design of Counterfort

The counterfort is designed as a cantilever beam whose depth is equal to the length of the heel slab at the base and reduces to the thickness of the stem at the top.

Maximum moment at the base of counterfort,

$$M_{max} = K_a.\gamma_e.h^3/6 \times L_e$$

Where,

L_e – c/c distance from counterfort

$$M_{max} = 1932.5 \text{ kNm}$$

Factored M_{max} = 2898.75 kNm

$$A_{st} = 2755.5 \text{ mm}^2$$

Assume 25 mm φ bar,

Number of bars required = 2755.5/491.5 = 5.61≈ 6.

The main reinforcement is provided along the slanting face of the counterfort.

Curtailment of Reinforcement

Not all the 6 bars need to be taken to the free end. Three bars are taken straight to the entire span of the beam. One bar is cut at a distance of,

$$\frac{n-1}{n} = \frac{h_1^2}{8.5^2}$$

Where,

n - The total number of bars.

h_1 - The distance from top.

When $n = 6$, $h_1 = 7.75$ m [from bottom].

The second part is cut at a distance of $\frac{n-2}{n} = \frac{h_2^{\,2}}{8.5^2}$ $h_2 = 6.94$ m [from bottom]

The third part is cut at a distance of,

$$\frac{n-3}{n} = \frac{h_3^{\,2}}{8.5^2}$$

$h_3 = 6.01$ m [from bottom]

Vertical ties and horizontal ties are provided to connect the counterfort with the vertical stem and the heel slab.

Design of Horizontal Ties

Closed stirrups are provided to the vertical stem and the counterfort. Considering 1m strip, the tension resisted by reinforcement is given by the lateral pressure on the wall multiplied by the contributing area.

$$T = C_a.\gamma_e.h \times h, \text{ where } A_{st} = \frac{T}{0.87\ fy}$$

$$T = 1/3 \times 16 \times (9 - 0.5) \times 3.54 = 160.48 \text{ kN}$$

Factored force, T = 1.5 × 160.48kN

$A_{st} = 666.72\ mm^2$. For 10mm φ, Spacing = 110mm.

Provide 10mm @ 110mm c/c closed stirrups as horizontal ties.

Design of Vertical Ties

The vertical stirrup connects the counterfort and the heel slab. Considering 1m strip, the tensile force is the product of the average downward pressure and the spacing between the counterfort.

$$T = \text{Avg}(43.56\ \&\ 107) \times L_e = 266.49 \text{ kN}$$

Factored T = 399.74kN

$A_{st} = 1107.15\ mm^2$. For 10 mm φ, Spacing = 70.93mm.

Provide 10 mm @ 70 mm c/c.

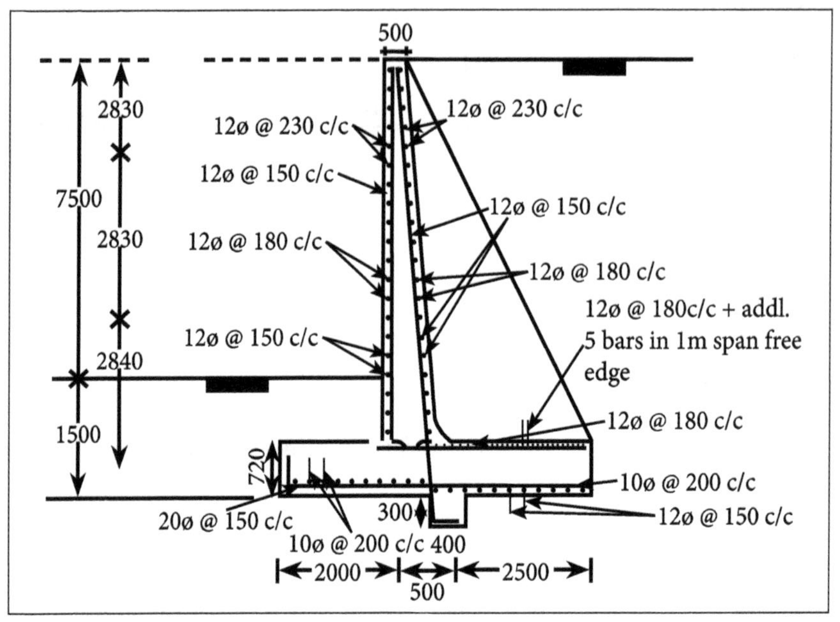

Reinforcement details of stem, toe slab and heel slab.

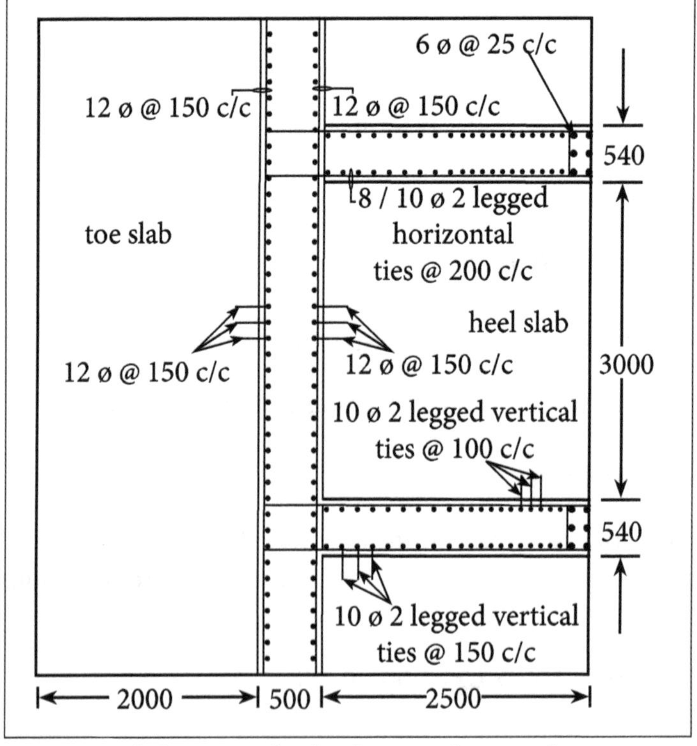

Reinforcement details of stem and counterfort.

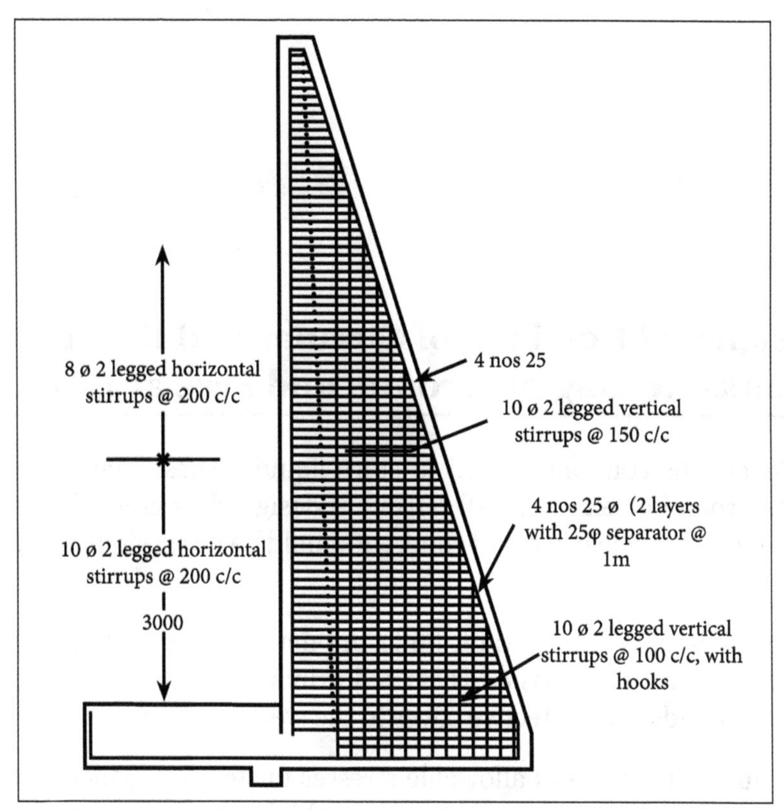

Section through counterfort showing counterfort reinforcement.

7

Design and Detailing of Water Tanks

7.1 Design and Detailing of Circular and Rectangular Water Tanks Resting on Ground and Free at Top

Storage tanks are the containers used to store liquids. These may be resting on the ground, underground or elevated. All tanks are designed as crack-free structures for durability and to prevent leakage and concrete should be impervious to eliminate seepage.

The reinforced concrete circular water tanks are mainly subjected to direct tension in the form of hoop force. This tension is carried primarily by the steel. Concrete is considered only to provide the protective cover to the steel reinforcement.

In such structures, the values of allowable stresses in steel and concrete are restricted so that the strains in steel and concrete are not high, consequently crack widths are limited. This provision minimizes the danger of corrosion of steel.

In other hand, in the tanks with fixed bases, the walls are subjected to bending tension. The members under direct tension are designed by elastic theory and those subjected to bending tension are designed either by elastic or limit states theory.

General Consideration

To achieve above objectives IS:3370 has recommended the following measures:

- Cement content: The concrete used for tanks should be a minimum of M25 grade mix so as to provide not only the strength but also higher density to prevent seepage. The cement content should not be less than 300 kg/m^3 to get water-tightness and not more than 530 kg/m^3 to avoid cracking due to shrinkage of concrete.

  A well-graded aggregate with a water-cement ratio less than 0.5 is recommended for making impervious concrete.

- Permissible stresses in steel: Plain mild or HYSD steel reinforcement can be used in the storage tanks. The permissible stress in the reinforcement is controlled by the strain and crack widths rather than by the strength. Deformed bars improve the level of the cracking strains in the concrete by even distribution and slip minimization.

In view of complexities associated with crack widths, a simplified approach through the reduced permissible stress in steel is recommended.

- Permissible stresses in concrete: To ensure untracked conditions, the permissible tensile stresses in concrete in reinforcement concrete members should not exceed the values, on the liquid-retaining face and also on the exterior face, for the members less than 225 mm thick.
- Cover to reinforcement: The cover to be provided on the surfaces not exposed to water is based on environmental conditions as stipulated in the code. The minimum clear or nominal cover to main reinforcement in direct tension shall be 20 mm or diameter of the bar, whichever is greater.

The minimum nominal cover is increased to 25 mm and 30 mm for the case of tension in bending and in the environment of alternate wetting and drying, respectively, but minimum cover should be 40 mm for the surfaces in contact with water. BS:5337, the British code on water retaining structures, has classified environmental exposure into the following three categories:

- Class A: Exposed to alternate wetting and drying (allowed crack width. 0.10 mm).
- Class B: Exposed to continuous contact with water. e.g. walls of liquid retaining structures (allowed crack width. 0.20 mm).
- Class C: Exposed only to outside air, i.e., to normal condition (allowed crack width. 0.30 mm).

The most common types of RC water tanks are as follows:

- Tanks resting on the ground.
- Underground tanks.
- Elevated water tanks on staging.

Water tightness is an important criterion in the construction of water tanks. Usually, richer mixes with M20 to M30 grade concrete are used. The permissible stresses in concrete under direct tension and bending are restricted to control cracking in concrete. There are three types of joints between the tank walls and floor:

- Flexible or free base.
- Fixed base.
- Hinged base.

In the case of a free or flexible base between the tank wall and base slab, the walls are free to slide, expand and the hoop tension developed in the circular walls due to hydrostatic pressure can be calculated easily.

However for hinged and fixed bases, the coefficients for moments and ring tension need to be considered. These coefficients are expressed as a function of the non-dimensional parameter.

$$(H^2 / Dt)$$

where,

H = Height of the tank.

D = Diameter of the tank.

t = Thickness of the tank wall.

Rectangular tanks are more economical than circular because the construction of circular tanks requires complicated and costly formwork. Moreover, compartmentation in a rectangular tank is much easier than the circular tanks.

The uses of rectangular tanks make the full use of the space available. The main components of a rectangular tank are side wall, base slab and roof slab.

Richer concrete mix of grades M20 to M30 are commonly used in the construction of water tanks. High quality concrete, in addition to providing water tightness, also has higher resistance to tensile stresses developed in the tank walls.

Minimum area of steel is 0.3% of gross area of section upto 100mm thick, reduced to 0.2% in section upto 450mm thick. For sections above 225mm thick, it provides two layers of reinforcement. The percentage of reinforcement in base or floor slab resisting directly on ground must not be less than 0.15% of the concrete section.

The minimum cover to all reinforcement should be not less than 25mm or the diameter of the bar whichever is greater.

Components of Water Tank

- Side Walls [Rectangular or cylindrical].
- Base slab.
- Cover slab or dome.
- Staging [Overhead] → Columns, Beams, Bracings.

Water Pressure Distribution

- On side walls, it acts linearly by varying from 0 at top to 'w_h' at bottom.

- On base slab, uniform pressure acts with intensity 'w_h'.

Permissible Stresses

The method of design adopted for design of water tank is working stress method. The permissible stress values in concrete and steel are as follows:

For $M_{20} \rightarrow \sigma_{cbc} = 7\ N/mm^2,\ \sigma_t = 5\ N/mm^2$
For Fe_{250} steel, $\sigma_{st} = 130 - 140\ N/mm^2$
For Fe_{415} steel, $\sigma_{st} = 230 - 240\ N/mm^2$

Permissible concrete stresses in calculations relating to resistance to cracking in water retaining structures.

Grade of concrete						
Stress (N/mm²)	M-15	M-20	M-25	M-30	M-35	M-40
Direct Tension	1.1	1.2	1.3	1.5	1.6	1.7
Bending Tension	1.5	1.7	1.8	2.0	2.2	2.4

Permissible stresses in steel reinforcement for strength calculations in Water retaining Structures.

Stress (N/mm²)	Plain Mild Steel bars	HYSD bars
Tensile stresses in members under direct tension	115	150
Tensile stresses of members in bending: On liquid retaining face of members.	115	150
On face away from liquid for members less than 225mm thick.	115	150
On face away from liquid for members 225mm or more in thickness.	125	190
Compressive stresses in columns subjected to direct load.	125	175

Circular Tanks Resting on Ground

The wall is designed for hoop stress, for which circumferential horizontal steel is provided. Minimum reinforcement is provided along the vertical direction.

Due to hydrostatic pressure, the tank has tendency to increase in diameter. This increase in diameter all along the height of the tank depends on the nature of joint at the junction of slab and wall as shown in the below figure.

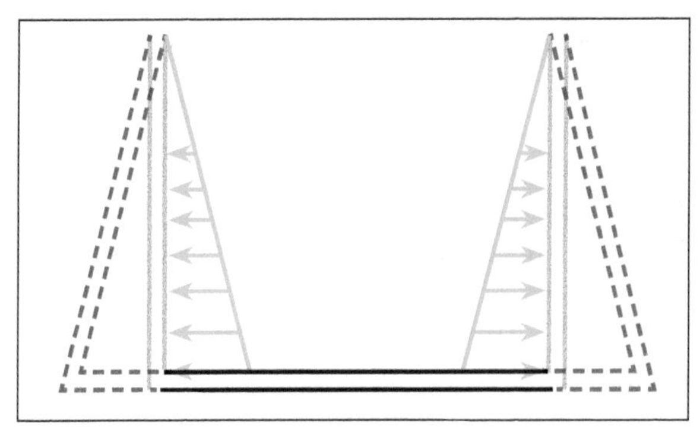

Tank with flexible base.

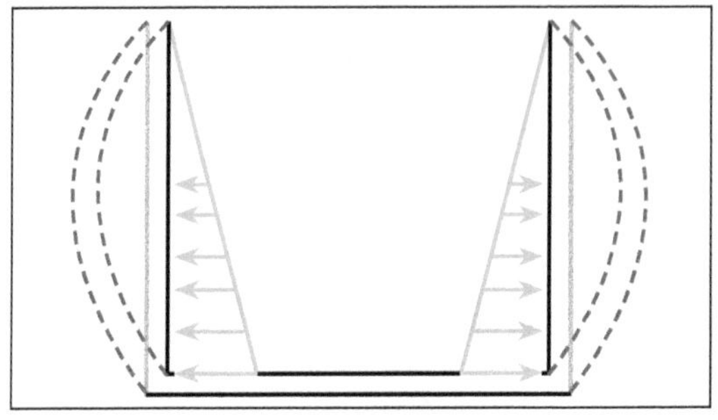

Tank with rigid base.

When the joints at base are flexible, hydrostatic pressure induces maximum increase in diameter at base and no increase in diameter at top. This is due to fact that hydrostatic pressure varies linearly from zero at top and maximum at base.

Deflected shape of the tank is shown in the figure. When the joint at base is rigid, the base does not move. The vertical wall deflects as shown in the above figure.

Design of Circular Tanks Resting on Ground with Flexible Base

Maximum hoop tension in the wall is developed at the base. This tensile force T is computed by considering the tank as thin cylinder.

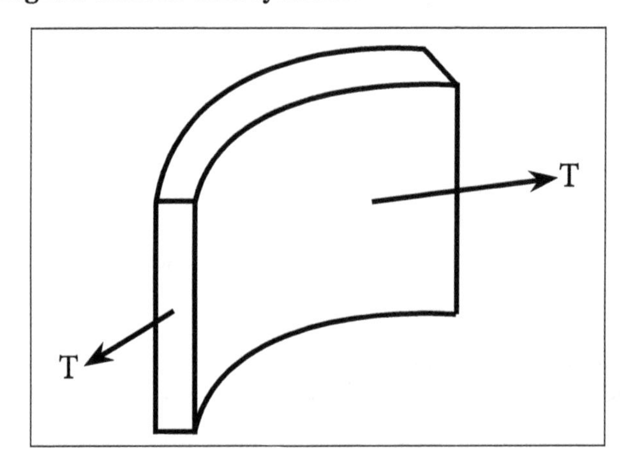

$$T = \gamma H \frac{D}{2};$$

Quantity of reinforcement required in form of hoop steel is computed as,

$$A_{st} = \frac{T}{\sigma_{st}} = \frac{\gamma HD/2}{\sigma_{st}} \text{ or } 0.3\% \text{ (minimum)}$$

When the thickness of the wall is less than 225 mm, the steel is placed at centre. When the thickness exceeds 225mm, at each face $A_{st}/2$ of steel as hoop reinforcement is provided.

In order to provide tensile stress in concrete to be less than permissible stress, the stress in concrete is computed using equation,

$$\sigma_c = \frac{T}{A_c + (m-1)A_{st}} = \frac{\gamma HD/2}{100t + (m-1)A_{st}} \text{ If } \sigma_c \leq \sigma_{cat}, \text{ where } \sigma_{cat} = 0.27\sqrt{f_{ck}}$$

Then the section is from cracking, otherwise the thickness has to be increased so that σ_c is less than σ_{cat}. While designing, the thickness of concrete wall can be estimated as $t = 30H + 50$ mm, where H is in meters.

Distribution steel in the form of vertical bars are provided such that minimum steel area requirement is satisfied. As base slab is resting on ground and no bending stresses are induced hence minimum steel distributed at bottom and the top are provided.

Design of Circular Tanks Resting on Ground with Rigid Base

Due to fixity at base of wall, the upper part of the wall will have hoop tension and lower part bend like cantilever. For shallow tanks with large diameter, hoop stresses are very small and the wall act more like cantilever.

For deep tanks of small diameter the cantilever action due to fixity at the base is small and the hoop action is predominant. The exact analysis of the tank to determine the portion of wall in which hoop tension is predominant and the other portion in which cantilever action is predominant, is difficult.

Simplified methods of analysis are:

- Reissner's method.
- Carpenter's simplified method.
- Approximate method.
- IS code method.

Use of IS code method for analysis and design of circular water tank with rigid base is given below.

Is Code Method

IS 3370 part IV gives coefficients for computing hoop tension, moment and shear for various values of H^2 / Dt.

Hoop tension, moment and shear is computed as,

$$T = \text{Coefficient}\ (\gamma_w HD/2)$$
$$M = \text{Coefficient}\ (\gamma_w H^3)$$
$$V = \text{Coefficient}\ (\gamma_w H^2)$$

Thickness of wall required is computed from BM consideration ie.,

$$d = \sqrt{\frac{M}{Qb}}$$

where,

$$Q = \tfrac{1}{2}\,\sigma_{cbc}\,jk$$

$$k = \frac{m\sigma_{cbc}}{m\sigma_{cbc} + \sigma_{st}}$$

$$j = 1 - (k/3)$$
$$b = 1000\text{mm}$$

Providing suitable cover, the overall thickness is then computed as t = d + cover.

Area of reinforcement in the form of vertical bars on water face is computed as,

$$\therefore A_{st} = \frac{M}{\sigma_{st}\,jd}$$

Area of hoop steel in the form of rings is computed as,

$$A_{st1} = \frac{T}{\sigma_{st}}$$

Distribution steel and vertical steel for outer face of wall is computed from minimum steel consideration.

Tensile stress computed from the following equation should be less than the permissible stress for safe design,

$$\sigma_s = \frac{T}{1000t + (m-1)A_{st}}$$ and the permissible stress is $0.27\sqrt{f_{ck}}$

Base slab thickness generally varies from 150mm to 250 mm and minimum steel is distributed to top and bottom of slab.

Rectangular Tank with Fixed Base Resting on Ground

Rectangular tanks are used when the storage capacity is small and circular tanks prove uneconomical for small capacity. Rectangular tanks should be preferably square in plan from point of view of economy. It is also desirable that longer side should not be greater than twice the smaller side.

Moments are caused in two directions of the wall, both in horizontal as well as in vertical direction. Exact analysis is difficult and such tanks are designed by approximate methods.

When the length of the wall is more in comparison to its height, the moments will be mainly in the vertical direction, the panel bends as vertical cantilever.

When the height is large in comparison to its length, the moments will be in the horizontal direction and panel bends as a thin slab supported on edges. For intermediate condition bending takes place both in horizontal and vertical direction.

In addition to the moments, the walls are also subjected to direct pull exerted by water pressure on some portion of walls. The walls are designed both for direct tension and bending moment.

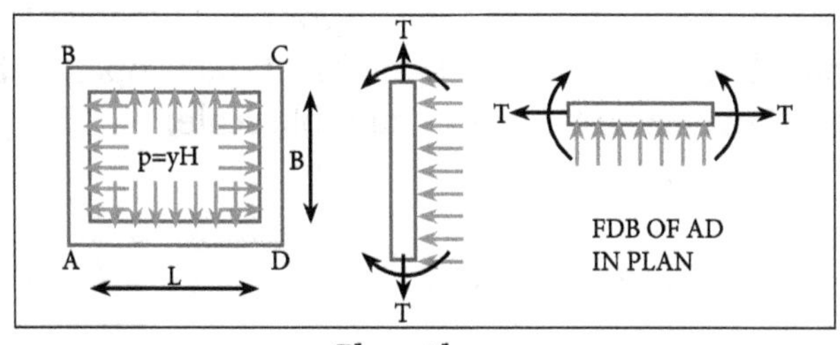

Plan at base.

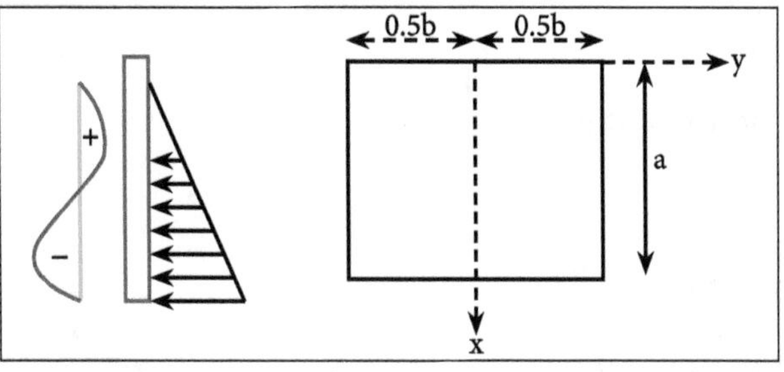

Bending moment diagram FBD of AB in plan.

IS3370 (Part-IV) gives tables for moments and shear forces in walls for certain edge condition. Table provides coefficient for max Bending moments in horizontal and vertical direction.

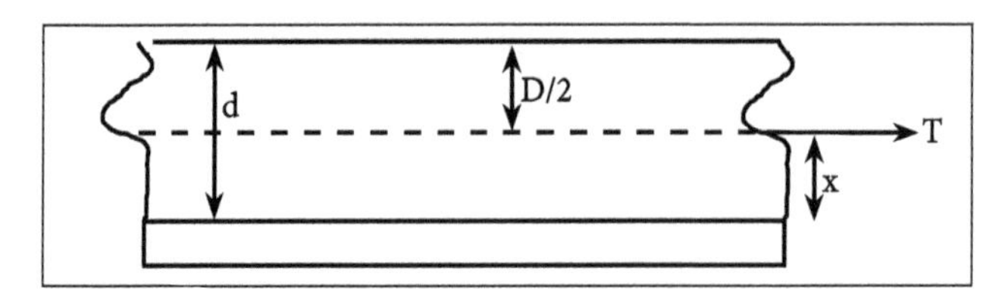

Maximum vertical moment $= M_x \gamma_w a^3$ (for $x/a = 1$, $y = 0$)

Maximum horizontal moment $= M_y \gamma_w a^3$ (for $x/a = 0$, $y = b/2$)

Tension in short wall is computed as $T_s = pL/2$

Tension in long wall $T_L = pB/2$

Horizontal steel is provided for net bending moment and direct tensile force,

$$A_{st} = A_{st1} + A_{st2};\ A_{st1} = \frac{M'}{\sigma_{st}\, jd};$$

M'=Maximum horizontal bending moment – T x; x = d-D/2

$$A_{st2} = T / \sigma_{st}$$

Problems

1. Let us design a circular water tank with flexible connection at base for a capacity of 4,00,000 liters. The tank rests on a firm level ground. The height of tank including a free board of 200 mm should not exceed 3.5m. The tank is open at top. Use M 20 concrete and Fe 415 steel. Let us also Draw to a suitable scale:

- Plan at base.
- Cross section through centre of tank.

Solution:

Given:

Capacity - 4,00,000 liters

M20, Fe415

Step 1: Dimension of tank.

Depth of water H=3.5 - 0.2 = 3.3 m

Volume $V = 4{,}00{,}000 / 1000 = 400\ m^3$

Area of tank $A = 400 / 3.3 = 121.2\ m^2$

Diameter of tank $D = \sqrt{\frac{4A}{\pi}}$

D = 12.42 m ≈ 13 m

The thickness is assumed as t = 30H + 50 = 149 ≈ 160 mm

Step 2: Design of Vertical wall.

Max hoop tension at bottom, $T = \gamma H \frac{D}{2} = \frac{10 \times 3.3 \times 13}{2} = 214.5\ kN$

Area of steel, $A_{st} = \frac{T}{\sigma_{st}} = \frac{T}{\sigma_{st}} = \frac{214.5 \times 10^3}{150} = 1430\ mm^2$

Minimum steel to be provided.

$A_{st,min} = 0.24\%$ of area of concrete $= 0.24 \times 1000 \times 160 / 100 = 384\ mm^2$

The steel required is more than the minimum required.

Let the diameter of the bar to be used be 16 mm, area of each bar $= 201\ mm^2$

Spacing of 16 mm diameter bar $= 1430 \times 1000 / 201 =$ 140.6 mm c/c

Provide #16 @ 140 c/c as hoop tension steel.

Step 3: Check for tensile stress.

Area of steel provided Ast, provided $= 201 \times 1000 / 140 = 1436.16\ mm^2$

Modular ratio, $m = \frac{280}{3\sigma_{cbc}} = \frac{280}{3 \times 7} = 13.33$

Stress in concrete, $\sigma_c = \frac{T}{1000\ t + (m-1) A_{st}} = \frac{214.5 \times 10^3}{1000 \times 160 + (13.33 - 1) 1436} = 1.2\ N/mm^2$

Permissible stress $\sigma_{cat} = 0.27 \sqrt{f_{ck}} =$ 1.2 N/mm^2

Actual stress is equal to permissible stress, hence safe.

Step 4: Curtailment of hoop steel.

Quantity of steel required at 1m, 2m and at top are tabulated. In this table the maximum spacing is taken a 3 × 160 = 480 mm.

Height from top	Hoop tension T = γHD/2 (kN)	$A_{st} = T / \sigma_{st}$	Spacing of #16 mm c/c
2.3 m	149.5	996	200
1.3 m	84.5	563.33	350
Top	0	Min steel (384 mm²)	400

Step 5: Vertical reinforcement.

For temperature and shrinkage distribution steel in the form of vertical reinforcement is provided @ 0.24 % i.e. A_{st} =384 mm².

Spacing of 10 mm diameter bar = 78.54 × 1000/384 = 204 mm c/c ≈200 mm c/c

Step 6: Tank floor.

As the slab rests on firm ground, minimum steel @ 0.3 % is provided. Thickness of slab is assumed as 150 mm. 8 mm diameter bars at 200 c/c is provided in both directions at bottom and top of the slab.

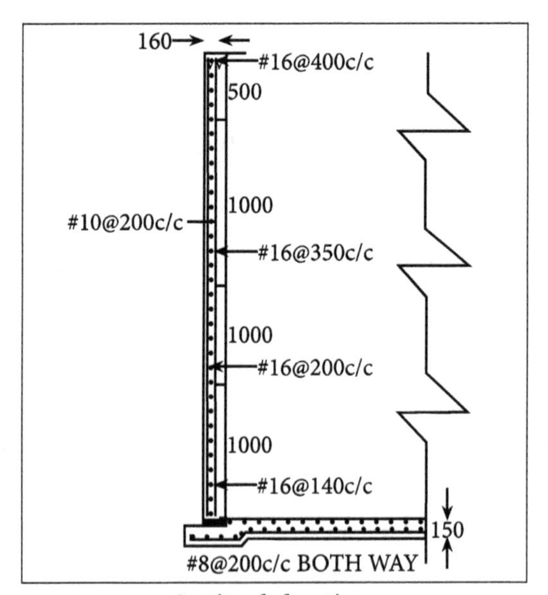

Sectional elevation.

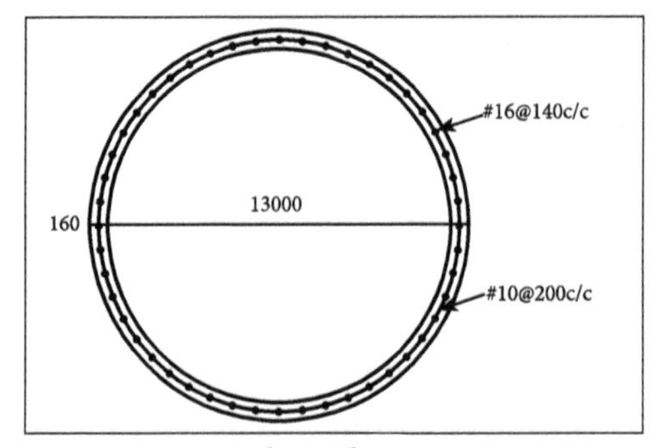

Plan at base.

2. A cylindrical tank of capacity 7,00,000 liters is resting on good unyielding ground. The depth of tank is limited to 5m. A free board of 300 mm may be provided. The wall and the base slab are cast integrally. Let us design the tank using M20 concrete and Fe415 grade steel and draw the following:

- Plan at base.
- Cross section through centre of tank.

Solution:

Given:

Capacity - 7,00,000 liters

Step 1: Dimension of tank.

H = 5-0.3 = 4.7 and volume $V = 700 \text{ m}^3$

$$A = 700 / 4.7 = 148.94 \text{ m}^2$$

$$D = \sqrt{(4 \times 148.94 / \pi)} = 13.77 \approx 14 \text{ m}$$

Step 2: Analysis for hoop tension and bending moment.

One meter width of the wall is considered and the thickness of the wall is estimated as t = 30H + 50 = 191 mm. The thickness of wall is assumed as 200 mm.

$$\frac{H^2}{Dt} = \frac{4.7^2}{14 \times 0.2} = 7.89 \approx 8$$

The maximum coefficient for hoop tension = 0.575.

$$T_{max} = 0.575 \times 10 \times 4.7 \times 7 = 189.175 \text{ kN}$$

The maximum coefficient for bending moment = -0.0146 (produces tension on water side).

$$M_{max} = 0.0146 \times 10 \times 4.7^3 = 15.15 \text{ kN} - \text{m}$$

Step 3: Design of section.

For M20 concrete $\sigma_{cbc} = 7$, For Fe415 steel $\sigma_{st} = 150$ MPa and m=13.33 for M20 concrete and Fe415 steel.

The design constants are:

$$k = \frac{m\sigma_{cbc}}{m\sigma_{cbc} + \sigma_{st}} = 0.39$$

$$j = 1 - (k/3) = 0.87$$

$$Q = \tfrac{1}{2}\,\sigma_{cbc}\,jk = 1.19$$

Effective depth is calculated as,

$$d = \sqrt{\frac{M}{Qb}} = \sqrt{\frac{15.15 \times 10^6}{1.19 \times 1000}} = 112.94 \text{ mm}$$

Let over all thickness be 200 mm with effective cover 33 mm $d_{provided} = 167$ mm

$$A_{st} = \frac{M}{\sigma_{st}\,jd} = \frac{15.15 \times 10^6}{150 \times 0.87 \times 167} = 695.16 \text{ mm}^2$$

Spacing of 16 mm diameter bar $= \dfrac{201 \times 1000}{695.16} = 289.23$ mmc/c (Max spacing 3d = 501mm)

Provide #16@275 c/c as vertical reinforcement on water face.

Hoop steel:

$$A_{st1} = \frac{T}{\sigma_{st}} = \frac{189.275 \times 10^3}{150} = 1261 \text{ mm}^2$$

Spacing of 12 mm diameter bar

$$= \frac{113 \times 1000}{1261} = 89.\text{ mmc/c}$$

Provide #12@80 c/c as hoop reinforcement on water face.

Actual area of steel provided,

$$A_{st} = \frac{113 \times 1000}{80} = 1412.5 \text{ mm}^2$$

Step 4: Check for tensile stress.

$$\sigma_c = \frac{T}{1000t + (m-1)A_{st}} = \frac{189.275 \times 10^3}{1000 \times 200 + (13.33 - 1) \times 1412.5} = 0.87 \text{ N/mm}^2$$

Permissible stress $= 0.27\sqrt{f_{ck}} = 1.2 \text{ N/mm}^2 > \sigma_c$ Safe.

Step 5: Distribution Steel.

Minimum area of steel is 0.24% of concrete area:

$$A_{st} = (0.24/100) \times 1000 \times 200 = 480 \text{ mm}^2$$

Spacing of 8 mm diameter bar $= \dfrac{50.24 \times 1000}{480} = 104.7.$ mmc/c

Provide #8 @ 100 c/c as vertical and horizontal distribution on the outer face.

Step 6: Base slab.

The thickness of base slab shall be 150 mm. The base slab rests on firm ground, hence only minimum reinforcement is provided.

$$A_{st} = (0.24/100) \times 1000 \times 150 = 360 \text{ mm}^2$$

Reinforcement for each face $= 180 \text{ mm}^2$

Spacing of 8 mm diameter bar $= \dfrac{50.24 \times 1000}{180} = 279.\text{mmc}/\text{c}$

Provide #8 @ 250 c/c as vertical and horizontal distribution on the outer face.

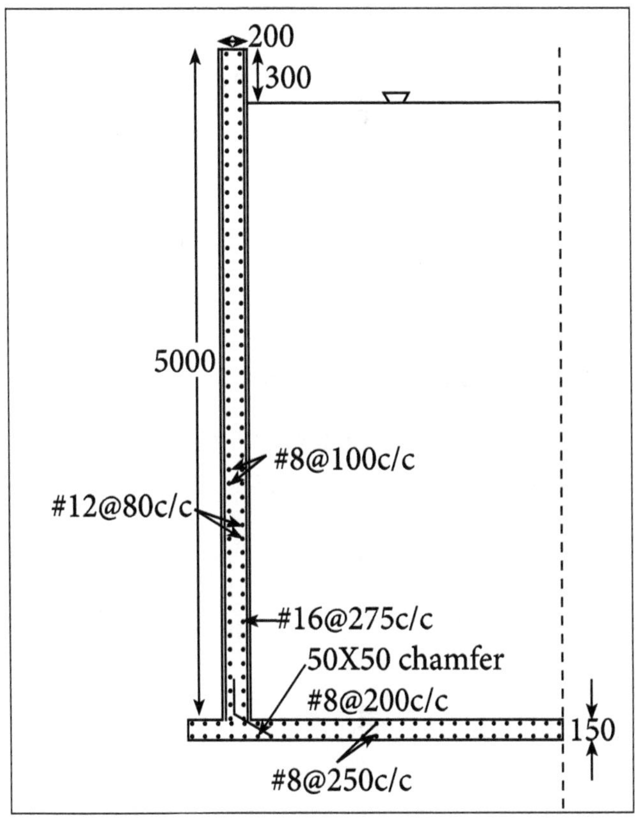

Sectional elevation.

3. Let us design a circular water tank with flexible base for a capacity of 4 lakh liters with the tank having a depth of 4m, including a free base of 200mm. Use M20 concrete and Fe415 steel.

Solution:

Given:

Circular Water Tank

Volume - 4 lakh liters

Depth - 4m

Area of water tank = Volume / Height

$$= 400/4 = 100\text{m}^2 \text{ [As Volume} = 4 \text{ Lakh Liters} = 400 \text{ m}^3]$$

$$\Pi.d^2/4 = 100 \text{ m}^2$$

$$d = 11.28\text{m}$$

Provide a diameter of 11.5m.

The height of water to be retained = 3.8m.

The wall is subjected to hoop tension acting along the circumferential direction. The hoop tension per meter height is given by,

$$\text{Hoop tension} = \frac{\gamma.h.D}{2} = (9.81 \times 3.8 \times 11.5)/2 = 214.34 \text{ kN}$$

Permissible stress in tension as per IS3370 for Fe415 steel is 150 N/mm².

$$A_{st,\ required} = 214.34 \times 10^3 / 150 = 1428.9 \text{ mm}^2$$

$$\text{Spacing} = 1000_{ast}/A_{st}, \rightarrow 16\text{mm @ 140mm c/c}$$

Thickness of tank is adopted based on the tensile stress concrete can take.

$$\sigma_{ct} = \frac{Ft}{A_c + mA_{st}}$$

$$A_c = 1000t,\ F = 214.34 \text{ kN},\ m = 280/3\,\sigma_{cbc} = 280/(3 \times 7) = 13.33 \approx 13$$

$$\sigma_{ct} = 1.2 = \frac{214.34}{1000t + (13.33 \times 1428.9)}$$

t = 160mm

Minimum thickness as per empirical formula is,

$$t_{min} = (30h + 50)\text{mm}$$

Where, $h \rightarrow m$

$$= 30 \times 3.8 + 50 = 164\text{mm}$$

Provide a thickness of 170mm.

Minimum reinforcement is provided as vertical steel.

Minimum A_{st} is 0.3% for 100mm section and 0.2% for 450mm section.

Therefore, for 170mm thickness, A_{st}, required is 0.28% of cross section.

$$A_{st} = (0.28/100) \times 1000 \times 170 = 476 \text{ mm}^2$$

Provide 8mm @ 100mm c/c.

Curtailment of Reinforcement

At 2m height,

$$\text{Hoop tension} = \frac{\gamma.h.D}{2} = (9.81 \times 1.8 \times 11.5)/2 = 101.5335 \text{ kN}$$

$$A_{st} = 101.535 \times 10^3 / 150 = 676.89 \text{ mm}^2$$

Provide 10mm @ 110mm c/c (or) 16mm @ 290mm c/c.

Design of Base Slab

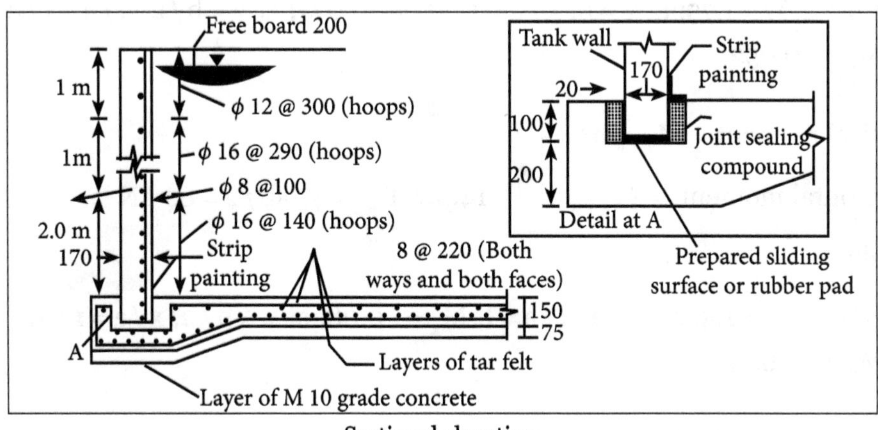

Sectional elevation.

Since the base slab rests directly on the ground, a nominal thickness of 150mm is provided and a minimum reinforcement of 0.3% of c/s is provided along both ways and along both the faces.

$$0.3\% \text{ of c/s} \rightarrow 0.3/100 \times 150 \times 1000 = 450 \text{ mm}^2, \text{ Required 8mm @110 mm c/c}$$

Provide 8mm @220 mm c/c on both the faces. Below the base slab, a layer of lean concrete mix M20 is provided for 75mm thickness with the layer of tar felt.

4. Let us design a rectangular water tank 5m × 4m with depth of storage 3m, resting on ground and whose walls are rigidly joined at vertical and horizontal edges. Assume M20 concrete and Fe415 grade steel. Sketch the details of reinforcement in the tank.

Solution:

Step 1: Analysis for moment and tensile force.

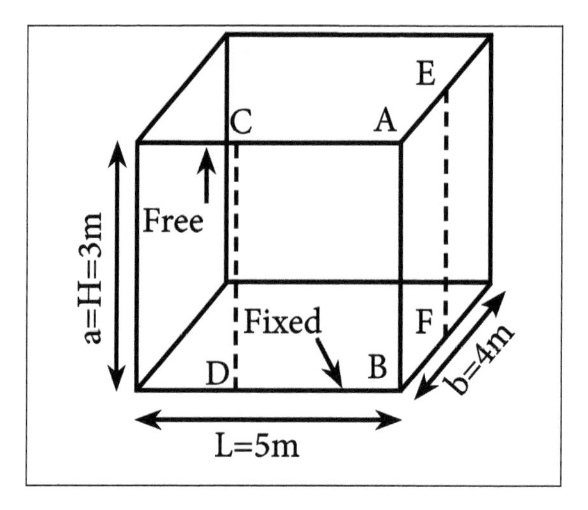

Long wall:

$L/a = 1.67 \approx 1.75$; at $y = 0$, $x/a = 1$, $M_x = -0.074$; at $y = b/2$, $x/a = 1/4$, $M_y = -0.052$

Max vertical moment $= M_x \gamma_w a^3 = -19.98$

Max horizontal moment $= M_y \gamma_w a^3 = -14.04$; $T_{long} = \gamma_w ab/2 = 60$ kN

Short wall:

$B/a = 1.33 \approx 1.5$; at $y = 0$, $x/a = 1$, $M_x = -0.06$; at $y = b/2$, $x/a = 1/4$, $M_y = -0.044$

Max vertical moment $= M_x \gamma_w a^3 = -16.2$

Max horizontal moment $= M_y \gamma_w a^3 = -11.88$; $T_{short} = \gamma_w aL/2 = 75$ kN

Step 2: Design constants.

$$\sigma_{cbc} = 7 \text{ MPa}, \ \sigma_{st} = 150 \text{ MPa}, \ m = 13.33$$

$$k = \frac{m\sigma_{cbc}}{m\sigma_{cbc} + \sigma_{st}} = 0.38$$

$$j = 1 - (k/3) = 0.87$$
$$Q = \tfrac{1}{2}\, \sigma_{cbc}\, jk = 1.15$$

Step 3: Design for vertical moment.

For vertical moment, the maximum bending moment from long and short wall,

$$(M_{max})_x = -19.98 \text{ kN} - \text{m}$$

$$d = \sqrt{\frac{m}{Qb}} = \sqrt{\frac{19.98 \times 10^6}{1.15 \times 1000}} = 131.8 \text{ mm}$$

Assuming effective cover as 33mm, the thickness of wall is,

$$t = 131.88 + 33 = 164.8 \text{ mm} \approx 170 \text{ mm}$$

$$d_{provided} = 170 - 33 = 137 \text{ mm}$$

$$A_{st} = \frac{M}{\sigma_{st}\, jd} = \frac{19.98 \times 10^6}{150 \times 0.87 \times 137} = 1117.54 \text{ mm}^2$$

Spacing of 12 mm diameter bar $= \dfrac{113 \times 1000}{1117.54} = 101.2$ mmc/c (Max spacing 3d=411mm)

Provide #12 @ 100 mm c/c.

Distribution steel

Minimum area of steel is 0.24% of concrete area,

$$A_{st} = (0.24/100) \times 1000 \times 170 = 408 \text{ mm}^2$$

Spacing of 8 mm diameter bar $= \dfrac{50.24 \times 1000}{408} = 123.19$ mmc/c

Provide #8 @ 120 c/c as distribution steel.

Provide #8 @ 120 c/c as vertical and horizontal distribution on the outer face.

Step 4: Design for Horizontal moment.

Horizontal moments at the corner in long and short wall produce unbalanced moment at the joint. This unbalanced moment has to be distributed to get balanced moment using moment distribution method.

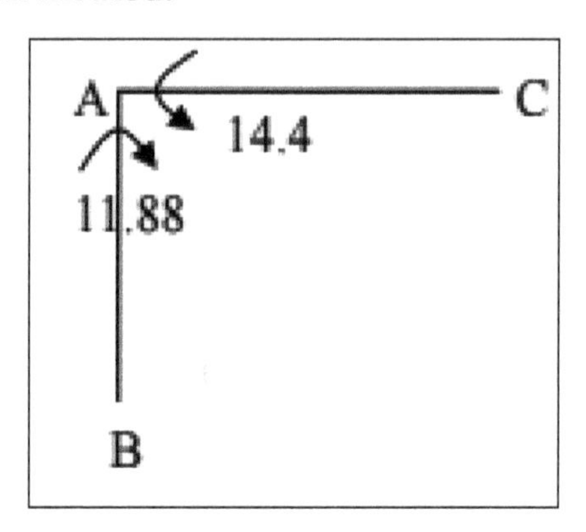

$$K_{AC} = \frac{1}{5};\ K_{AC} = \frac{1}{5};\ \sum K = \frac{9}{20}$$

$$DF_{AC} = \frac{1/5}{9/20} = 0.44$$

$$DF_{AB} = \frac{1/4}{9/20} = 0.56$$

Table: Moment distribution.

Joint	A	
Member	AC	AB
DF	0.44	0.56
FEM	-14	11.88
Distribution	0.9328	1.1872
Final Moment	-13.0672	13.0672

The tension in the wall is computed by considering the section at height H_1 from the base. Where, H_1 is greater of i) H/4, ii) 1m, i.e., i) 3/4 = 0.75, ii) 1m; $H_1 = 1\text{m}$

Depth of water $h = H - H_1 = 3 - 1 - 2\text{m};\ p = \gamma_w h = 10 \times 2 = 20\ \text{kN/m}^2$

Tension in short wall $T_s = pL/2 = 50\ \text{kN}$

Tension in long wall $T_L = pB/2 = 40\ \text{kN}$

Net bending moment $M' = M - Tx$, where, $x = d - D/2 = 137 - (170/2) = 52\text{mm}$

$$M' = 13.0672 - 50 \times 0.052 = 10.4672 \text{ kN} - \text{m}$$

$$A_{st1} = \frac{10.4672 \times 10^6}{150 \times 0.87 \times 137} = 585.46 \text{ mm}^2$$

$$A_{st2} = \frac{50 \times 10^3}{150} = 333.33 \text{ mm}^2$$

$$A_{st} = A_{st1} + A_{st2} = 918.79 \text{ mm}^2$$

Spacing of 12 mm diameter bar:

$$= \frac{113 \times 1000}{918.74} = 123 \text{ mmc/c (Max spacing 3d=411mm)}$$

Provide #12@120 mm c/c at corners.

Step 5: Base Slab.

The slab is resting on firm ground. Hence nominal thickness and reinforcement is provided. The thickness of slab is assumed to be 200 mm and 0.24% reinforcement is provided in the form of #8 @ 200 c/c at top and bottom. A haunch of 150 × 150 × 150 mm size is provided at all corners.

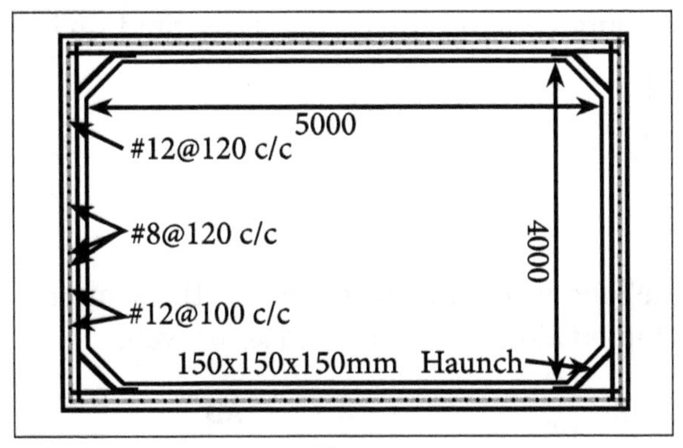

Plan.

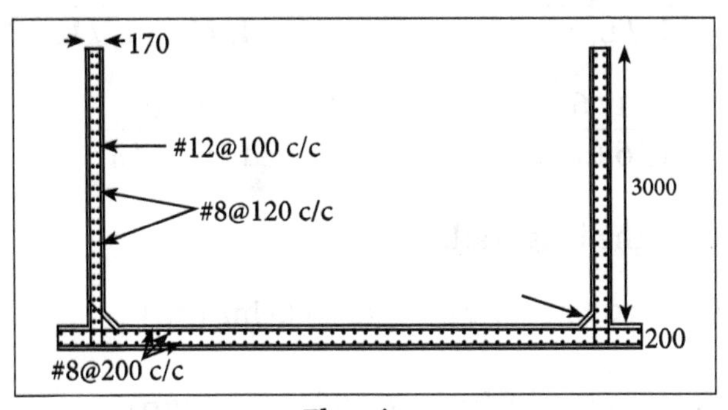

Elevation.

5. Let us design a rectangular tank of size 4m × 6m with height 3m. The tank rests on firm ground. Use M20 concrete and Fe415 steel. Take design constants j = 0.853 & R = 1.32.

Solution:

Given:

Rectangular tank

Size: 4m × 6m

Height: 3m

$j = 0.853$

$R = 1.32$

Pressure exerted by water, $p = w\,(H - h)$

Where, h = 1m or H/4 → 1m or 0.75m = 1 m [greater]

$$P = 9.81\,(3-1) = 19.62 \text{ kN/m}^2$$

To find the final moment at the junction of long wall and short wall based on the fixed end moment and distribution factor, the moment distribution is done.

$$\text{D.F.} = \frac{I_1/L_1}{I_1/L_1 + I_2/L_2}$$

Joint A:

The stiffness along the long wall and short wall are the same $(I_1 = I_2)$, since uniform thickness of wall is adopted along long wall and short wall.

	AB	AD
D.F	$\dfrac{I_1/L_1}{I_1/L_1 + I_2/L_2}$	$\dfrac{I_2/L_2}{I_1/L_1 + I_2/L_2}$
	$\dfrac{1/6}{1/6+1/4}$	$\dfrac{1/4}{1/6 \;\; 1/4}$

Short wall is stiffer than the long wall.

$$M_{FAB} = pL^2/12 = 19.62 \times 6^2/12 = 58.86 \text{ kNm } (3p)$$

$$M_{FAD} = pB^2/12 = 19.62 \times 4^2/12 = 26.16 \text{ kNm } (1.33p)$$

	AB	AD
FEM	3p	-1.33p
DF	0.4	0.6
BM	(-1.67 × 0.4)=-0.67p	(-1.67 × 0.6) = -1.002p
CO	-	-
Mf	+2.33p	-2.33p

$$M_f = 2.33p = 2.33 \times 19.62 = 45.72 \text{ kNm}$$

Moment At Midspan

Long wall:

$$\frac{pL^2}{8} - M_f = 42.57 \text{ kNm}$$

Short wall:

$$\frac{pB^2}{8} - M_f = -6.48 \text{ kNm}$$

The reinforcement is provided for maximum moment generated. Therefore, maximum moment generated in the water tank is 45.72 kNm.

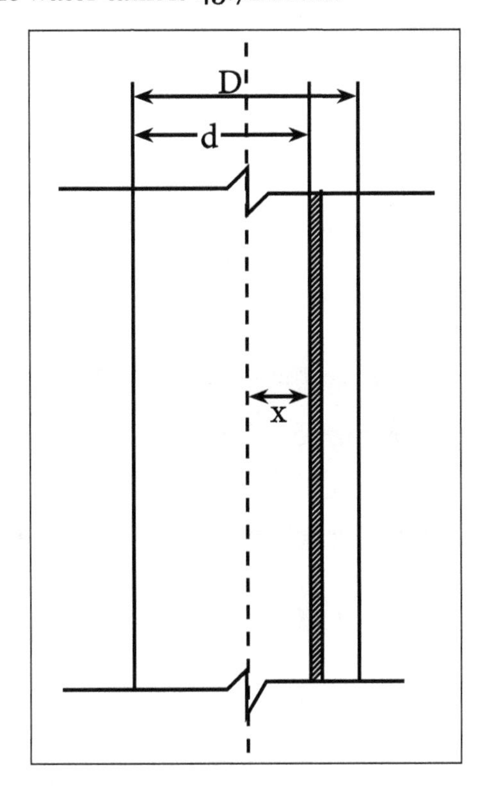

$$A_{stL} = \frac{M - P_L x}{\sigma_{st} . j.d} + \frac{P_L}{\sigma_{st}}$$

$$A_{stB} = \frac{M - P_B x}{\sigma_{st} . j.d} + \frac{P_B}{\sigma_{st}}$$

$$P_L = p \times B/2 = 19.62 \times 4/2 = 39.24 \text{ kN}$$
$$P_B = p \times L/2 = 19.62 \times 6/2 = 58.86 \text{ kN}$$

$$d_{req} = \sqrt{\frac{M}{Rb}} = \sqrt{\frac{45.72 \times 10^6}{1.32 \times 1000}} = 186 \text{ mm}$$

It provides d = 190 mm, D = 220 mm, effective depth of 190mm and effective cover of 30 mm.

$$x = (D/2) - \text{Effective Cover}$$
$$= 220/2 - 30$$
$$= 80 \text{ mm}$$

Area of steel required for tension in the wall along the longer side and shorter side,

$$A_{stL} = \frac{45.72 \times 10^6 - (39.24 \times 80 \times 10^3)}{150 \times 0.853 \times 190} + \frac{39.24 \times 10^3}{150}$$

$$A_{stL} = 2013 \text{ mm}^2$$

$$A_{stL} = \frac{45.72 \times 10^6 - (39.24 \times 80 \times 10^3)}{150 \times 0.853 \times 190} + \frac{39.24 \times 10^3}{150}$$

$$A_{stL} = 2079 \text{ mm}^2$$

Provide 20mm φ bar spacing required along horizontal direction is 150mm.

Provide 20mm @ 300mm c/c along both faces in the horizontal direction along short wall and long wall.

The vertical cantilever moment for a height of 'h' $= w.H.h^2/6 \; [L/B < 2]$

$$= 9.81 \times 3 \times (12/6)$$
$$= 4.905 \text{ kNm}$$

$$A_{st} = \frac{M}{\sigma_{st} . j.d} = 201.76 \text{ mm}^2$$

Minimum A_{st} = 0.3% of c/s

$$= 0.3/100 \times 1000 \times 220$$
$$= 660 \text{ mm}^2$$

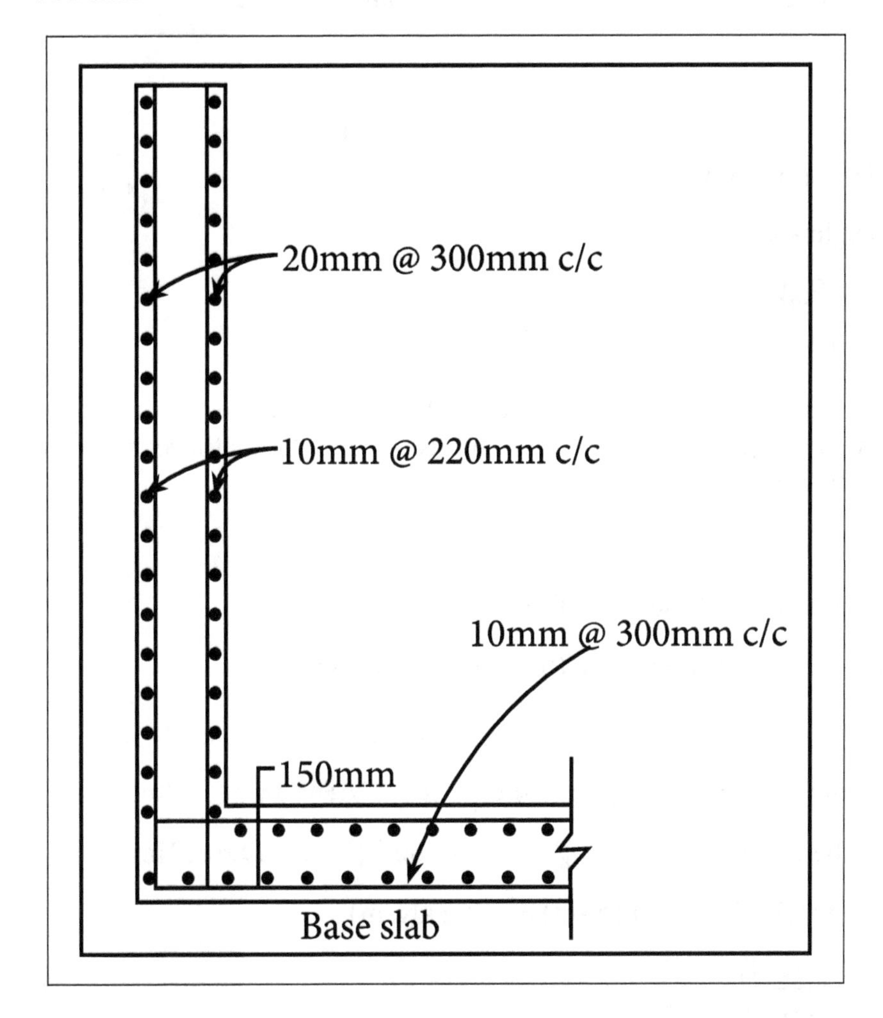

Spacing is provided for the maximum of the above two [66 mm²].

Required 10 mm @ 110 mm c/c.

Provide 10 mm @ 220 mm c/c as vertical reinforcement along both the faces. For the base slab, provide a nominal thickness of 150 mm and minimum A_{st} of 0.3% of c/s.

$$A_{st} = 0.3/100 \times 1000 \times 150 = 450 \text{ mm}^2$$

Spacing of 10mm bars required = 170 mm.

Provide 10 mm @ 300 mm c/c along both faces both ways.

Provide 75 mm lean mix with a layer of tar felt which acts as a water bar provided between the tank and lean mix concrete.

6. Let us design a water tank of size 4m x 9m with height 3m. Use M20 concrete and Fe415 steel. The design constants are j = 0.853 and R = 1.32.

Solution:

Given:

Size - 4m × 9m

Height - 3m

j = 0.853

R = 1.32

Since, L/B > 2, the tank behaves such that the long wall acts as a cantilever member with moment w.H3/6 and short wall is subjected to both cantilever moment and horizontal bending moment.

Long wall:

Cantilever moment at base $= w.H^3/6$

Where, h = 1m or 3/4m = 1m = 9.81 × 33 / 6 = 44.145 kNm

Short wall:

Cantilever moment $= w.H.h^2/12 = 9.81 \times 3 \times 1^2/2 = 14.715$ kNm

Horizontal bending moment $= p_B{}^2/16 = 19.62 \times 4^2/16 = 19.62$ kNm

(where, p = w (H – h) = 9.81 (3 – 1) = 19.62 kN/m)

$$A_{stL} = \frac{M - P_L x}{\sigma_{st} . j.d} + \frac{P_L}{\sigma_{st}}$$

Maximum of the three moments = 44.145 kNm

Where, $P_L = p \times B/2$

$= 19.62 \times (4/2)$

$= 39.24$ KN

$P_B = p = 19.62$ KN

$$d_{req} = \sqrt{\frac{M}{R.b}} = \sqrt{\frac{44.145 \times 10^6}{1.32 \times 1000}} = 182.87 \text{ mm}$$

Provide d = 190 mm, D = 220 mm

Provide effective depth of 190 mm and effective cover of 30 mm.

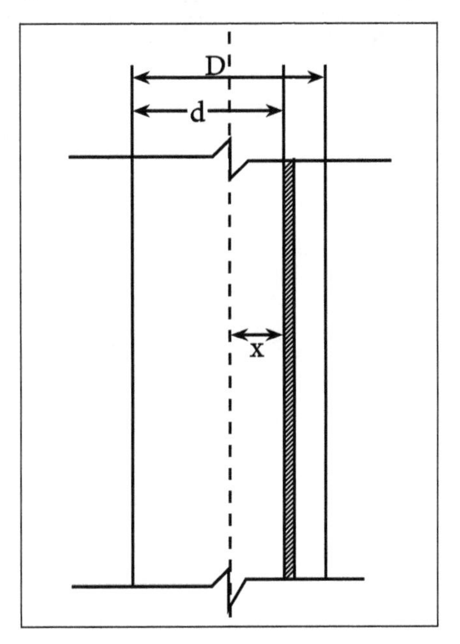

x = D /2 – Eff. Cover

= 220 / 2 – 30

= 80 mm

Area of steel required for Tension in the wall along the longer side,

$$A_{stL} = \frac{44.145 \times 10^6 - (39.24 \times 80 \times 10^3)}{150 \times 0.853 \times 190} + \frac{39.24 \times 10^3}{150} = 1948.35 \text{ mm}^2$$

Provide 20mm φ bar, spacing required along horizontal direction is 160mm.

Provide 20mm @ 300mm c/c along both faces in the horizontal direction, along short wall and long wall.

Along the Shorter Side

Along Vertical Direction:

$$A_{stB} = \frac{14.72 \times 10^6 - (19.62 \times 80 \times 10^3)}{150 \times 0.853 \times 180} + \frac{19.62 \times 10^3}{150} = 690.45 \text{ mm}^2$$

Required - 10mm @ 220mm c/c on both faces.

Along Horizontal Direction:

$$A_{stB} = \frac{19.62 \times 10^6 - \left(19.62 \times 80 \times 10^3\right)}{150 \times 0.853 \times 180} + \frac{19.62 \times 10^3}{150} = 897.5 \text{ mm}^2$$

Required - 10mm @160mm c/c on both faces.

Distribution steel

100mm $\rightarrow$ 0.3%

450mm $\rightarrow$ 0.2%

220mm $\rightarrow$ 0.26%

Minimum $A_{st} = 0.26\%$ of $c/s = 0.26 / 100 \times 1000 \times 220 = 572 \text{ mm}^2$

Minimum Required - 10mm @ 260mm c/c, which is less than the above two values.

Base Slab

Since base slab is resting directly on the ground, nominal thickness of 150mm is provided and minimum reinforcement of 0.3% of cross-section is provided both ways along both faces.

$$A_{st,\ required} = 0.3\% \text{ of } c/s = 465 \text{ mm}^2.$$

Provide 8mm @ 200mm c/c.

Below the base slab, a layer of lean concrete mix is provided with 75mm thick tar felt layer between them.

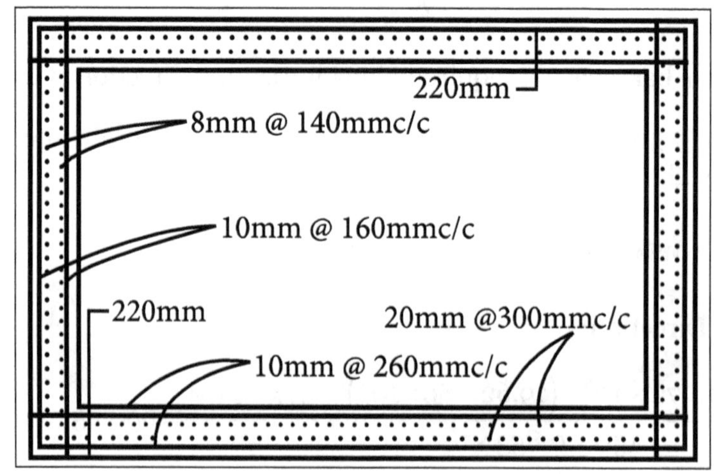

Reinforcement detail.

8

Design and Detailing of Portal Frames

8.1 Design and Detailing of Simple Portal Frames Subjected to Gravity Loads

Portal frames are generally low-rise structures, comprising columns and horizontal or pitched rafters, connected by moment-resisting connections.

Resistance to lateral and vertical actions is provided by the rigidity of the connections and the bending stiffness of the members, which is increased by a suitable haunch or deepening of the after sections. This form of continuous frame structure is stable in its plane and provides a clear span that is unobstructed by bracing.

Portal frames are very common, in fact 50 percentage of constructional steel used in the UK is in portal frame construction. They are very efficient for enclosing large volumes, therefore they are often used for industrial storage, retail and commercial applications as well as for agricultural purposes.

Anatomy of a Typical Portal Frame

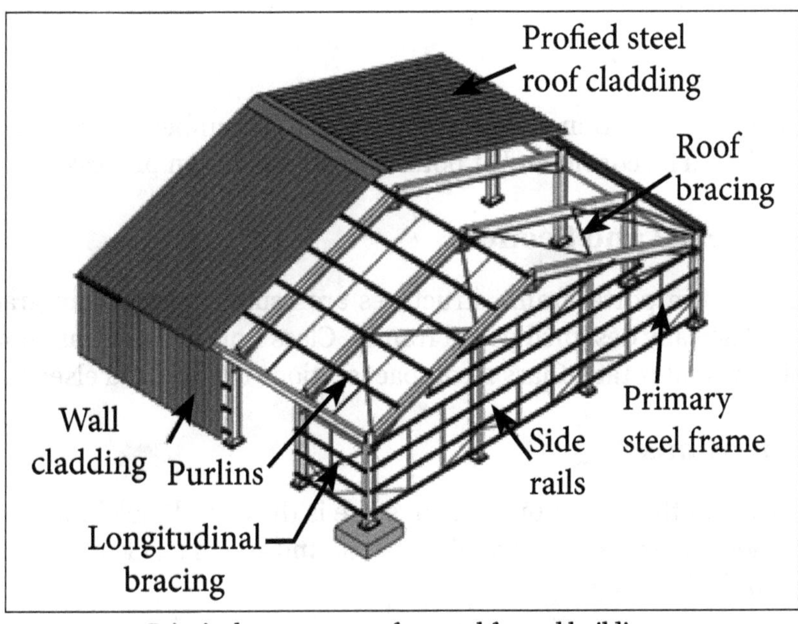

Principal components of a portal framed building.

A portal frame building comprises a series of transverse frames braced longitudinally. The primary steelwork consists of columns and rafters, which form portal frames and bracing. The end frame can be either a portal frame or a braced arrangement of columns and rafters.

The light gauge secondary steelwork consists of side rails for walls and purlins for the roof. The secondary steelwork supports the building envelope, but also plays an important role in restraining the primary steelwork.

The roof and wall cladding separate the enclosed space from the external environment as well as providing thermal and acoustic insulation. The structural role of the cladding is to transfer loads to secondary steelwork and also to restrain the flange of the purlin or rail to which it is attached.

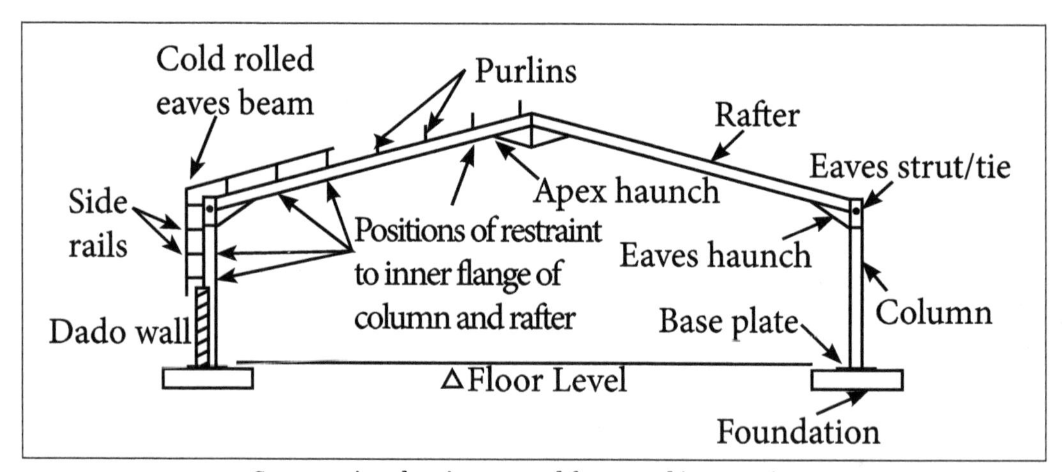

Cross-section showing a portal frame and its restraints.

Design Considerations

In the design and construction of any structure, a large number of inter-related design requirements should be considered at both stage in the design process.

Choice of Material and Section

Steel sections used in portal frame structures are usually specified in grade S275 or S355 steel. In plastically designed portal frames, Class 1 plastic sections should be used at hinge positions that rotate, Class 2 compact sections can be used elsewhere.

Frame Dimensions

A critical decision at the conceptual design stage is the overall height and width of the frame, to give adequate clear internal dimensions and adequate clearance for the internal functions of the building.

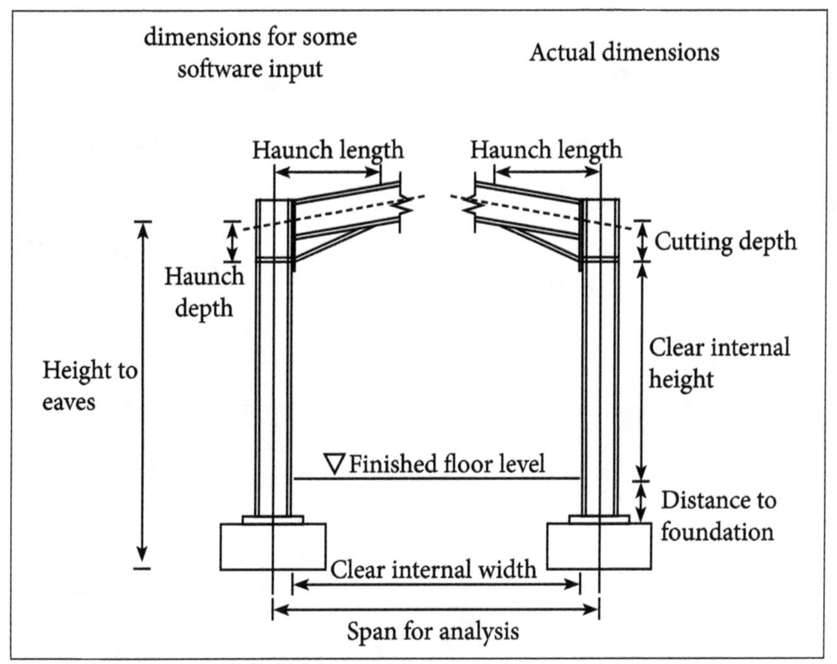

Dimensions used for analysis and clear internal dimensions.

Clear Span and Height

The clear span and height required by the client are key to determining the dimensions to be used in the design and should be established early in the design process.

The client requirement is likely to be the clear distance between the flanges of the two columns the span will therefore be larger, by the section depth. Any requirement for brickwork or blockwork around the columns should be established as this may affect the design span.

Where a clear internal height is specified, this will usually be measured from the finished floor level to the underside of the haunch or suspended ceiling if present.

Main Frame

The main (portal) frames are generally fabricated from UB sections with a substantial eaves haunch section, which may be cut from a rolled section or fabricated from plate. A typical frame is characterized by:

- A span between 15m and 50 m.
- An clear height (from the top of the floor to the underside of the haunch) between 5 and 12 m.
- A roof pitch between 5° and 10° (6° is commonly adopted).

- A frame spacing between 6 and 8 m.
- Haunches in the rafters at the eaves and apex.
- A stiffness ratio between the column and rafter section of approximately 1.5.
- Light gauge purlins and side rails.
- Light gauge diagonal ties from some purlins and side rails to restrain the inside flange of the frame at certain locations.

Haunch Dimensions

Typical haunch with restraints.

The use of a haunch at the eaves reduces the required depth of rafter by increasing the moment resistance of the member where the applied moments are highest. The haunch also adds stiffness to the frame, reducing deflections and facilitates an efficient bolted moment connection.

The eaves haunch is typically cut from the same size rolled section as the rafter or one slightly larger and is welded to the underside of the rafter. The length of the eaves haunch is generally 10% of the frame span.

The haunch length generally means that the hogging moment at the end of the haunch is approximately equal to the largest sagging moment close to the apex. The depth from the rafter axis to the underside of the haunch is approximately 2% of the span.

The apex haunch may be cut from a rolled section often from the same size as the rafter or fabricated from plate. The apex haunch is not usually modeled. In the frame analysis and is only used to facilitate a bolted connection.

Positions of Restraints

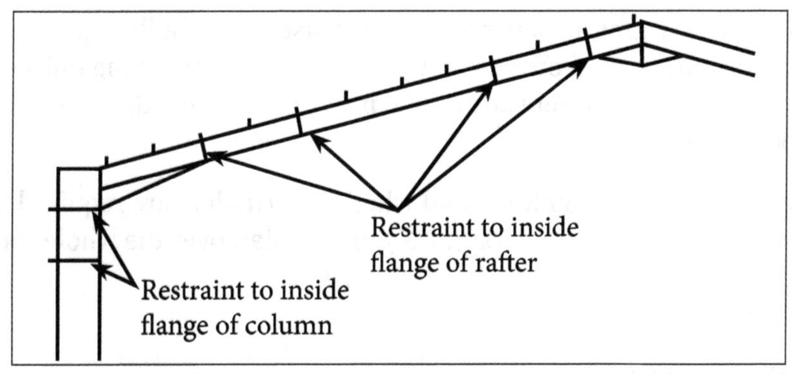

General arrangement of restraints to the inside flange.

During initial design the rafter members are normally selected according to their cross sectional resistance to bending moment and axial force. In later design stages stability against buckling needs to be verified and restraints positioned judiciously.

The buckling resistance is likely to be more significant in the selection of a column size, as there is usually less freedom to position rails to suit the design requirements, rail position may be dictated by doors or windows in the elevation.

If introducing intermediate lateral restraints to the column is not possible, the buckling resistance will determine the initial section size selection. It is therefore essential to recognise at this early stage if the side rails may be used to provide restraint to the columns.

Only continuous side rails are effective in providing restraint. Side rails interrupted by (for example) roller shutter doors, cannot be relied on as providing adequate restraint.

Where the compression flange of the rafter or column is not restrained by purlins and side rails, restraint can be provided at specified locations by column and rafter stays.

Actions

Advice on actions can be found in BS EN 1991 and on the combinations of actions in BS EN 1990. It is important to refer to the UK National Annex for the relevant Euro-code part for the structures to be constructed in the UK.

Permanent Actions

Permanent actions are the self-weight of the structure, secondary steelwork and cladding. Where possible, unit weights of materials should be obtained from manufacturers' data. Where information is not available, these may be determined from the data in BS EN 1991-1-1.

Service Loads

Service loads will vary greatly depending on the use of the building. In portal frames heavy point loads may occur from suspended walkways, air handling units etc. It is necessary to consider carefully where additional provision is needed, as particular items of plant must be treated individually.

Depending on the use of the building and whether sprinklers are required, it is normal to assume a service loading of 0.1–0.25 kN/m^2 on plan over the whole roof area.

Variable Actions

Imposed Roof Loads

Imposed loads on roofs are given in the UK NA to BS EN 1991-1-1 and depend on the roof slope. A point load, q_k is given, which is used for local checking of roof materials and fixings and a uniformly distributed load, q_k, to be applied vertically. The loading for roofs not accessible except for normal maintenance and repair is given in the below table.

It should be noted that imposed loads on roofs should not be combined with either snow or wind.

Imposed Loads on Roofs

Roof slope, α	qk (kN/m²)
α < 30°	0.6
30° < α < 60°	0.6[60 - α)/30]
α > 60°	0

Snow Loads

Snow loads may sometimes be the dominant gravity loading. Their value should be determined from BS EN 1991-1-3 and its UK National Annex.

Any drift condition must be allowed for not only in the design of the frame itself, but also in the design of the purlins that support the roof cladding. The intensity of loading at the position of maximum drift often exceeds the basic of minimum uniform snow load.

The calculation of drift loading and associated purlin design has been made easier by the major purlin manufacturers, most of whom offer free software to facilitate rapid design.

Wind Actions

Wind actions in the UK should be determined using BS EN 1991-1-4 and its UK National Annex. This Eurocode gives much scope for national adjustment and therefore its annex is a substantial document.

Wind actions are inherently complex and likely to influence the final design of most buildings. The designer needs to make a careful choice between a fully rigorous, complex assessment of wind actions and the use of simplifications which ease the design process but make the loads more conservative. The free software for establishing wind pressures is available from purlin manufacturers.

Crane Actions

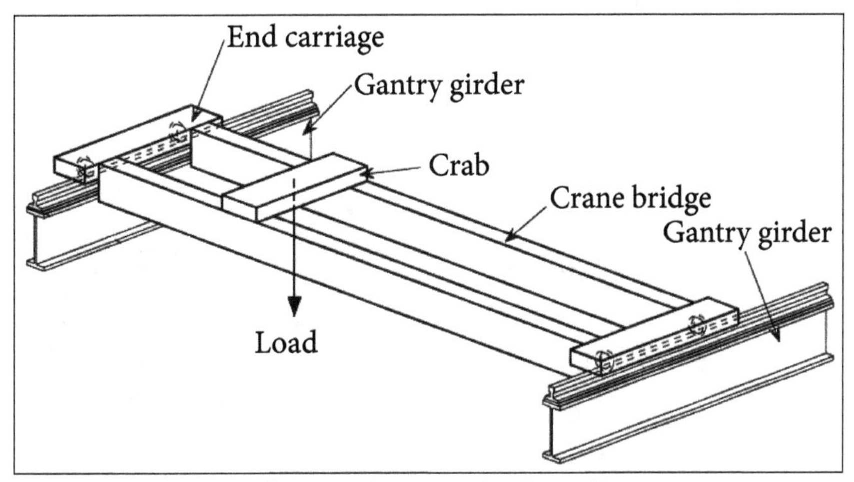

Gantry girders carrying an overhead traveling crane.

The most common form of cranage is the overhead type running on beams supported by the columns. The beams are carried on cantilever brackets or in heavier cases, by providing dual columns.

In addition to the self-weight of the cranes and their loads, the effects of acceleration and deceleration have to be considered. For simple cranes, this is by a quasi-static approach with amplified loads.

For heavy, high-speed or multiple cranes the allowances should be specially calculated with reference to the manufacturer.

Accidental Actions

The common design situations which are treated as accidental design situations are:

- Drifted snow, determined using Annex B of BS EN 1991-1-3.
- The opening of a dominant opening which was assumed to be shut at ULS.

- Each project should be individually assessed whether any other accidental actions are likely to act on the structure.

Robustness

Robustness requirements are designed to ensure that any structural collapse is not disproportionate to the cause. BS EN 1990 sets the requirement to design and construct robust buildings in order to avoid disproportionate collapse under accidental design situations. BS EN 1991-1-7 gives details of how this requirement should be met.

For more information on robustness refer to SCI P391.

For many portal frame structures no special provisions are needed to satisfy robustness requirements set by the Eurocode.

Fire

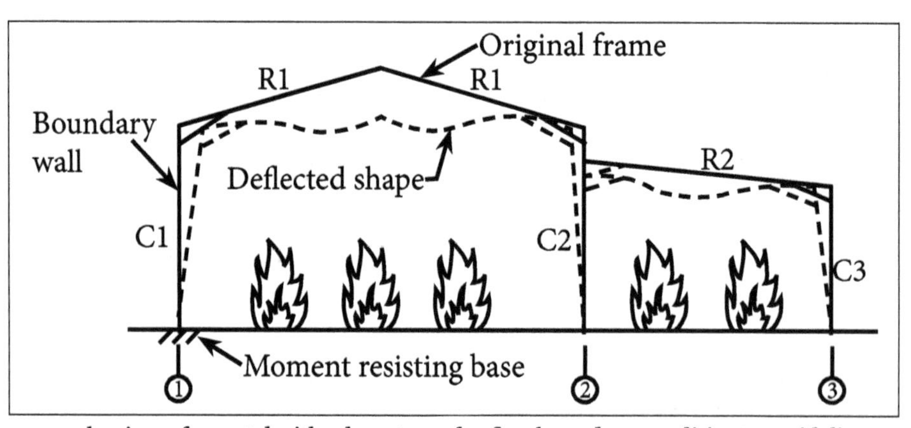

Collapse mechanism of a portal with a lean-to under fire, boundary condition on gridelins 2 and 3.

In the United Kingdom, structural steel in single storey buildings does not normally require fire resistance. The most common situation in which it is required to fire protect the structural steelwork is where prevention of fire spread to adjacent buildings, a boundary condition, is required.

There are a low number of other, rare, instances, for example when demanded by an insurance provider, where structural fire protection may be required.

When a portal frame is close to the boundary, there are many requirements aimed at stopping fire spread by keeping the boundary intact:

- The use of fire resistant cladding.
- The provision of a moment resisting base (as it is assumed that in the fire condition rafters go into catenary).
- Application of fire protection of the steel up to the underside of the haunch.

Combinations of Actions

BS EN 1990 gives rules for establishing combinations of actions, with the values of relevant factors given in the UK National Annex. BS EN 1990 covers both ultimate limit state (ULS) and serviceability limit state (SLS), although for the SLS, onward reference is made to the material codes (for example BS EN 1993-1-1 for steelwork) to identify which expression should be used and what SLS limits should be observed.

All combinations of actions that can occur together should be considered, however if certain actions cannot be applied simultaneously, they should not be combined.

Guidance on the application of Eurocode rules on combinations of actions can be found in SCI P362 and, specifically for portal frames, in SCI P399.

Frame Analysis at ULS

At the ultimate limit state (ULS), the methods of frame analysis fall broadly into two types, elastic analysis and plastic analysis.

Plastic Analysis

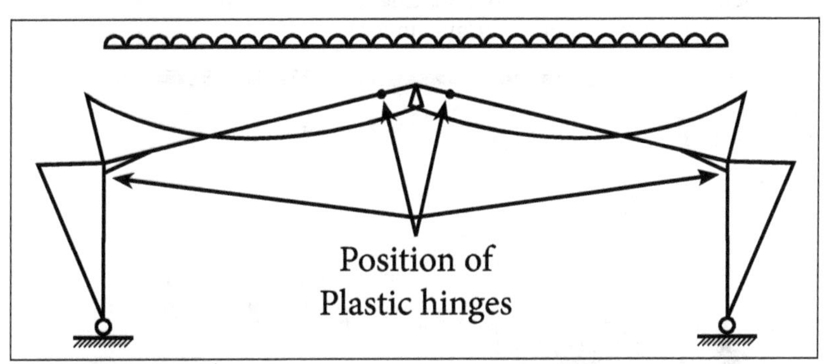

Bending moment diagram resulting from the plastic analysis of a symmetrical portal frame under symmetrical loading.

The term plastic analysis is used to cover both rigid-plastic and elastic-plastic analysis. Plastic analysis commonly results in a more economical frame because it allows relatively large redistribution of bending moments throughout the frame, due to plastic hinge rotations.

These plastic hinge rotations occur at sections where the bending moment reaches the plastic moment or resistance of the cross-section at loads below the full ULS loading.

The rotations are normally considered to be localized at "plastic hinges" and allow the capacity of underutilized parts of the frame to be mobilized. The figure shows typical positions where plastic hinges form in a portal frame. Two hinges lead to a collapse, but in the illustrated example, due to symmetry, designers need to consider all possible hinge locations.

Elastic Analysis

A typical bending moment diagram resulting from an elastic analysis of a frame with pinned bases is shown the figure below. In case, the maximum moment (at the eaves) is higher than that calculated from a plastic analysis. Each column and haunch has to be designed for these large bending moments.

Where, deflections (SLS) govern design, there may be no advantage in using plastic analysis for the ULS. If stiffer sections are selected in order to control deflections, it is quite possible that no plastic hinges form and the frame remains elastic at ULS.

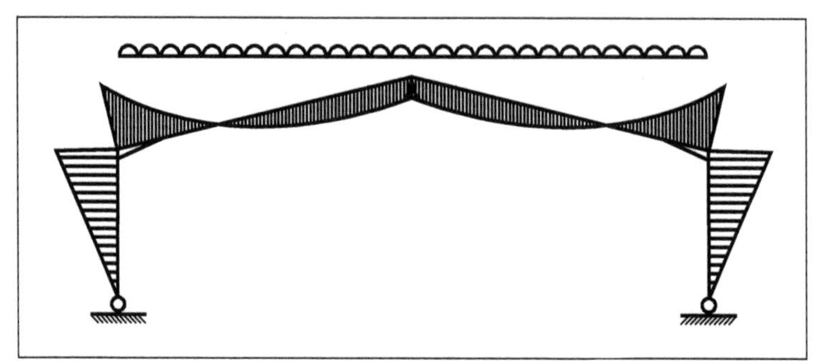

Bending moment diagram resulting from the elastic analysis of a symmetrical portal frame under symmetrical loading.

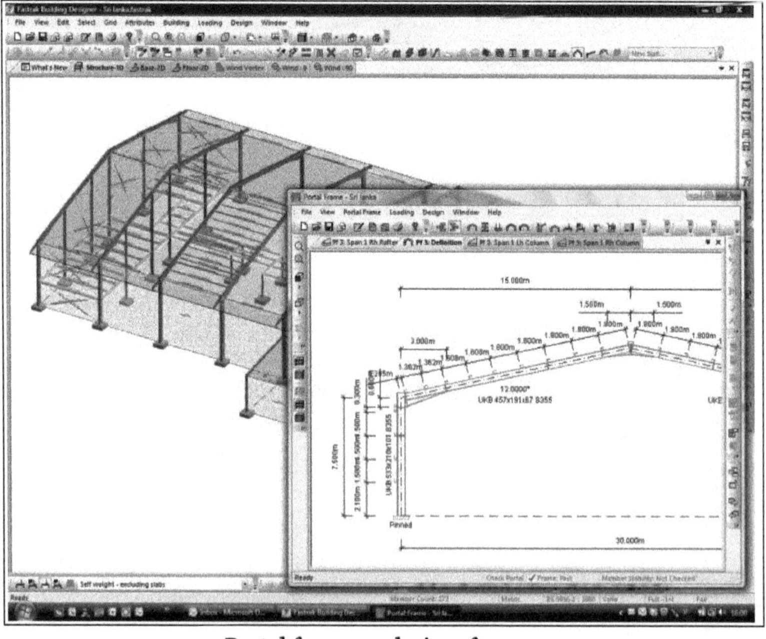

Portal frame analysis software.

In-plane Frame Stability

When any frame is loaded, it deflects and its shape under load is various from the un-deformed shape.

The deflection has a number of effects:

- The vertical loads are eccentric to the bases, which leads to further deflection.
- The apex drops, reducing the arching action.
- Applied moments curve members, Axial compression in curved members causes increased curvature (which may be perceived as a reduced stiffness).
- Taken together, these effects mean that a frame is less stable (nearer collapse) than a first-order analysis suggests. The objective of assessing frame stability is to determine if the difference is significant.

Second Order Effects

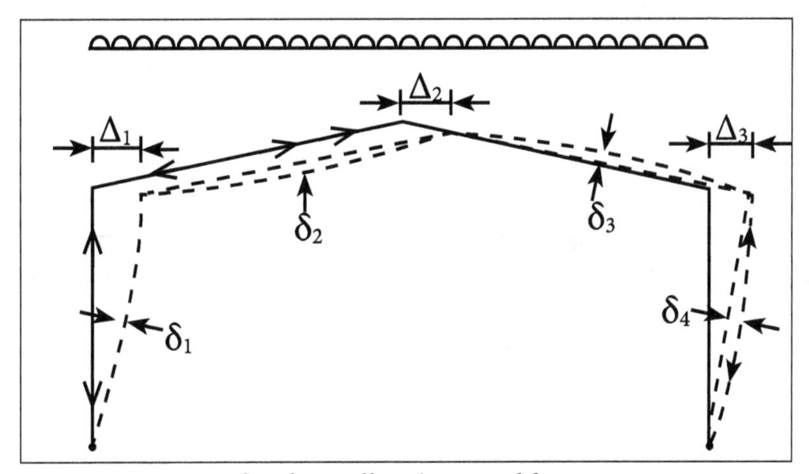

P-δ and P-Δ effects in a portal frame.

The geometrical effects described above are second-order effects and should not be confused with nonlinear behaviour of materials. As shown in the figure above there are two categories of second-order effects:

- Effects of displacements of the intersections of members, usually called P-Δ effects. BS EN 1993-1-1 describes this as the effect of deformed geometry.
- Effects of deflections within the length of members, usually called P-δ effects.

Second-order analysis is the term used to describe analysis methods in which the effects of increasing deflection under increasing load is considered explicitly in the solution, so that the results include the P-δ and P-Δ effects.

8.1.1 First-order and Second-order Analysis

For either plastic analysis of frames or elastic analysis of frames, the choice of first-order analysis or second-order analysis depends on the in plane flexibility of the frame, characterize by the calculation of the α_{cr} factor.

Calculation of á $_{cr}$

The effects of the deformed geometry (P-Δ effects) are assessed in BS EN 1993–1–1 by calculating the factor α_{cr}, defined as,

$$\alpha_{cr} = \frac{F_{cr}}{F_{Ed}} \geq 10$$

Where,

F_{Ed} – The design load on the structure.

F_{cr} – The elastic critical buckling load for global instability mode, based on initial elastic stiffnesses.

α_{cr} may be found using software or using an approximation as long as the frame meets certain geometric limits and the axial force in the rafter is not 'significant'. Rules are given in the Eurocode to identify when the axial force is significant.

When the frame falls outside the specified limits, as is the case for very many orthodox frames, the simplified expression cannot be used. In these circumstances, an alternative expression may be used to calculate an approximate value of α_{cr}, referred to as $\alpha_{cr,est}$.

Sensitivity to Effects of the Deformed Geometry

The limitations to the use of first-order analysis are defined in BS EN 1993–1–1, Section 5.2.1 (3) and the UK National Annex Section NA.2.9 as:

For elastic analysis: $\alpha_{cr} \geq 10$

For plastic analysis:

$\alpha_{cr} \geq 5$ for combinations with gravity loading with frame imperfections, provided that:

- The span, L, does not exceed 5 times the mean height of the columns.
- h_r satisfies the criterion: $(h_r / s_a)^2 + (h_r / s_b)^2 \leq 0.5$ in which s_a and s_b are the horizontal distances from the apex to the columns. For a symmetrical frame this expression simplifies to $h_r \leq 0.25L$.

$\alpha_{cr} \geq 10$ for combinations with gravity loading with frame imperfections for clad structures provided that the stiffening effects of masonry infill wall panels or diaphragms of profiled steel sheeting are not taken into account.

Design

Once the analysis has been completed, allowing for second-order effects if necessary, the frame members must be verified.

Both the cross-sectional resistance and the buckling resistance of the members must be verified. In-plane buckling of members need not be verified as the global analysis is considered to account for all significant in-plane effects. SCI P399 identifies the likely critical zones for member verification. SCI P397 contains numerical examples of member verifications.

Cross-section Resistance

Member bending, axial and shear resistances must be verified. If the shear or axial force is high, the bending resistance is reduced so combined shear force and bending and axial force and bending resistances need to be verified.

In typical portal frames neither the shear force nor the axial load is sufficiently high to reduce the bending resistance. When the portal frame forms the chord of the bracing system, the axial load in the rafter may be significant and this combination of actions should be verified.

Although all cross-sections need to be verified, the likely key points are at the positions of maximum bending moment:

- In the column at the underside of the haunch.
- In the rafter at the sharp end of the haunch.
- In the rafter at the maximum sagging location adjacent to the apex.

Member Stability

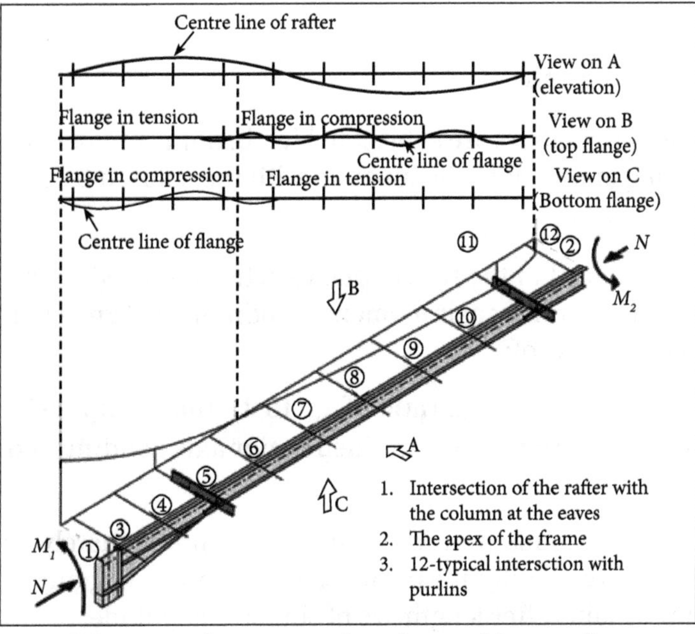

Diagrammatic representation of a portal frame rafter.

The figure shows a diagrammatic representation of the issues that need to be addressed when considering the stability of a member within a portal frame, in this example a rafter between the eaves and apex.

The following points should be noted:

- Purlins provide intermediate lateral restraint to one flange. Depending on the bending moment diagram this may be either the tension or compression flange.
- Restraints to the inside flange can be provided at purlin positions, producing a torsional restraint at that location.

In-plane, no member buckling checks are required, as the global analysis has accounted for all significant in-plane effects. The analysis has accounted for any significant second-order effects and frame imperfections are usually accounted for by including the equivalent horizontal force in the analysis. The effects of in-plane member imperfections are small enough to be ignored.

$$\frac{N_{Ed}}{\frac{\chi_z N_{Rk}}{\gamma_{M1}}} + k_{zy}\frac{M_{y,Ed} + \Delta M_{y,Ed}}{\chi_{LT}\frac{M_{y,Rk}}{\gamma_{M1}}} + k_{zz}\frac{M_{z,Ed} + \Delta M_{z,Ed}}{\frac{M_{z,Rk}}{\gamma_{M1}}} \leq 1 \quad \ldots(1)$$

Because there are no minor axis moments in a portal frame rafter, Expression (1) simplifies to:

$$\frac{N_{Ed}}{N_{b,z,Rd}} + k_{zy}\frac{M_{y,Ed}}{M_{b,Rd}} \leq 1.0$$

Rafter Design and Stability

In the plane of the frame rafters are subject to high bending moments, which vary from a maximum ‘hogging’ moment at the junction with the column to a minimum sagging moment close to the apex.

Compression is introduced in the rafters due to actions applied to the frame. The rafters are not subject to any minor axis moments. Optimum design of portal frame rafters is generally achieved by use of:

- A cross section with a high ratio of I_{yy} to I_{zz} that complies with the requirements of Class 1 or 2 under combined major axis bending and axial compression.
- A haunch that extends from the column for approximately 10% of the frame span. This will generally mean that the maximum hogging and sagging moments in the plain rafter length are of similar magnitude.

Out-of-Plane Stability

Purlins attached to the top flange of the rafter provide stability to the member in a number of ways:

- Direct lateral restraint, when the outer flange is in compression.
- Intermediate lateral restraint to the tension flange between torsional restraints, when the outer flange is in tension.
- Torsional and lateral restraint to the rafter when the purlin is attached to the tension flange and used in conjunction with rafter stays to the compression flange.
- Initially, the out-of-plane checks are completed to ensure that the restraints are located at appropriate positions and spacing.

Gravity Combination of Actions

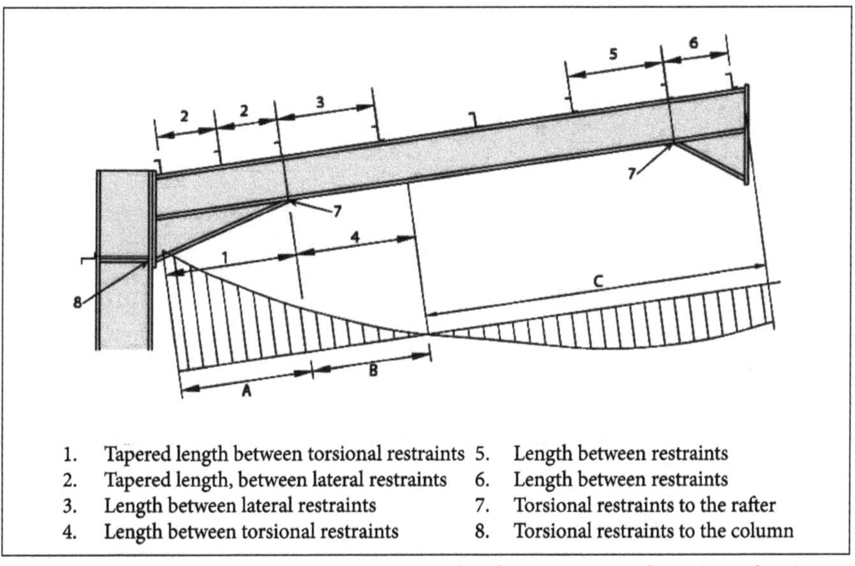

Typical purlin and rafter stay arrangement for the gravity combination of actions.

The figure shows a typical moment distribution for the gravity combination of actions, typical purlin and restraint positions as well as stability zones, which are referred to further.

Purlins are generally placed at up to 1.8 m spacing but this spacing may need to be reduced in the high moment regions near the eaves.

In Zone A, the bottom flange of the haunch is in compression. The stability checks are complicated by the variation in geometry along the haunch. The bottom flange is partially or wholly in compression over the length of Zone B. In Zone C, the purlins provide lateral restraint to the top (compression) flange.

The selection of the appropriate check depends on the presence of a plastic hinge, the shape of the bending moment diagram and the geometry of the section (three flanges or two flanges). The objective of the checks is to provide sufficient restraints to ensure the rafter is stable out-of-plane.

Guidance on details of the out-of plane stability verification can be found in SCI P399.

The Uplift Condition

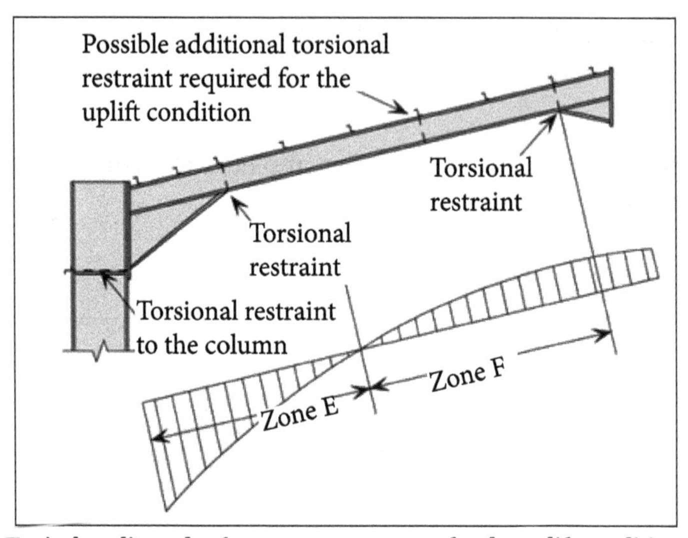

Typical purlin and rafter stay arrangement for the uplift condition.

In the uplift condition the top flange of the haunch will be in compression and will be restrained by the purlins. The moments and axial forces are smaller than those in the gravity load combination.

As the haunch is stable in the gravity combination of actions, it will certainly be so in the uplift condition, being restrained at least as well and under reduced loads.

In Zone F, the purlins will not restrain the bottom flange, which is in compression.

The rafter must be verified between torsional restraints. A torsional restraint will generally be provided adjacent to the apex. The rafter may be stable between this point and the virtual restraint at the point of contraflexure, as the moments are generally modest in the uplift combination.

If the rafter is not stable over this length, additional torsional restraints should be introduced and each length of the rafter verified.

In Plane Stability

No in-plane checks of rafters are required, as all significant in-plane effects have been accounted for in the global analysis.

Column Design and Stability

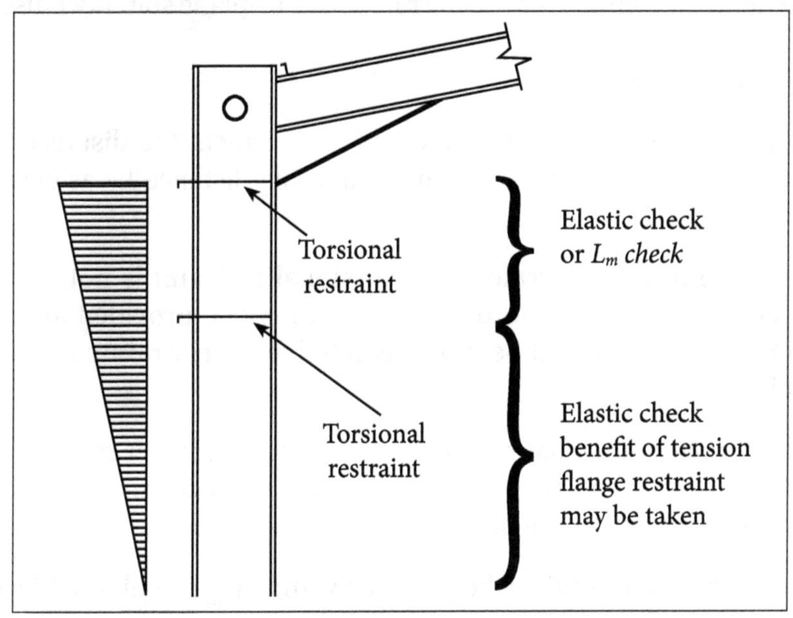

Typical portal frame column with plastic hinge at underside of haunch.

The most heavily loaded region of the rafter is reinforced by the haunch. By contrast, the column is subject to a similar bending moment at the underside of the haunch, but without any additional strengthening.

The optimum design for most columns is usually achieved by the use of:

- A cross section with a high ratio of I_{yy} to I_{zz} that complies with Class 1 or Class 2 under combined major axis bending and axial compression.
- A plastic section modulus that is approximately 50% greater than that of the rafter.

The column size will generally be determined at the preliminary design stage on the basis of the required bending and compression resistances.

Whether the frame is designed plastically or elastically, a torsional restraint should always be provided at the underside of the haunch. This may be from a side rail positioned at that level or by some other means.

Additional torsional restraints may be required between the underside of the haunch and the column base because the side rails are attached to the (outer) tension flange; unless restraints are provided the inner compression flange is unrestrained.

A side rail that is not continuous (for example, interrupted by industrial doors) cannot be relied upon to provide adequate restraint. The column section may need to be increased if intermediate restraints to the compression flange cannot be provided.

The presence of a plastic hinge will depend on loading, geometry and choice of column and rafter sections. In a similar way to the rafter, out-of-plane stability must be verified.

Out-of-plane Stability

If there is a plastic hinge at the underside of the haunch, the distance to the adjacent torsional restraint must be less than the limiting distance L_m as given by BS EN 1993-1-1Clause BB.3.1.1.

It may be possible to demonstrate that a torsional restraint is not required at the side rail immediately adjacent to the hinge, but may be provided at some greater distance. In this case there will be intermediate lateral restraints between the torsional restraints.

If the stability between torsional restraints cannot be verified, it may be necessary to introduce additional torsional restraints. If it is not possible to provide additional intermediate restraints, the size of the member must be increased.

In all cases, a lateral restraint must be provided within L_m of a plastic hinge.

When the frame is subject to uplift, the column moment will reverse. The bending moments will generally be significantly smaller than those under gravity loading combinations and the column is likely to remain elastic.

In Plane Stability

No in-plane checks of columns are required, as all significant in-plane effects have been accounted for in the global analysis.

Bracing

Bracing is required to resist longitudinal actions due to wind and cranes and to provide restraint to members.

It is common to use hollow sections as bracing members.

Vertical bracing

The primary functions of vertical bracing in the side walls of the frame are:

- To transmit the horizontal loads to the ground. The horizontal forces include forces from wind and cranes.
- To provide a rigid framework to which side rails and cladding may be attached so that the rails can in turn provide stability to the columns.
- To provide temporary stability during erection.

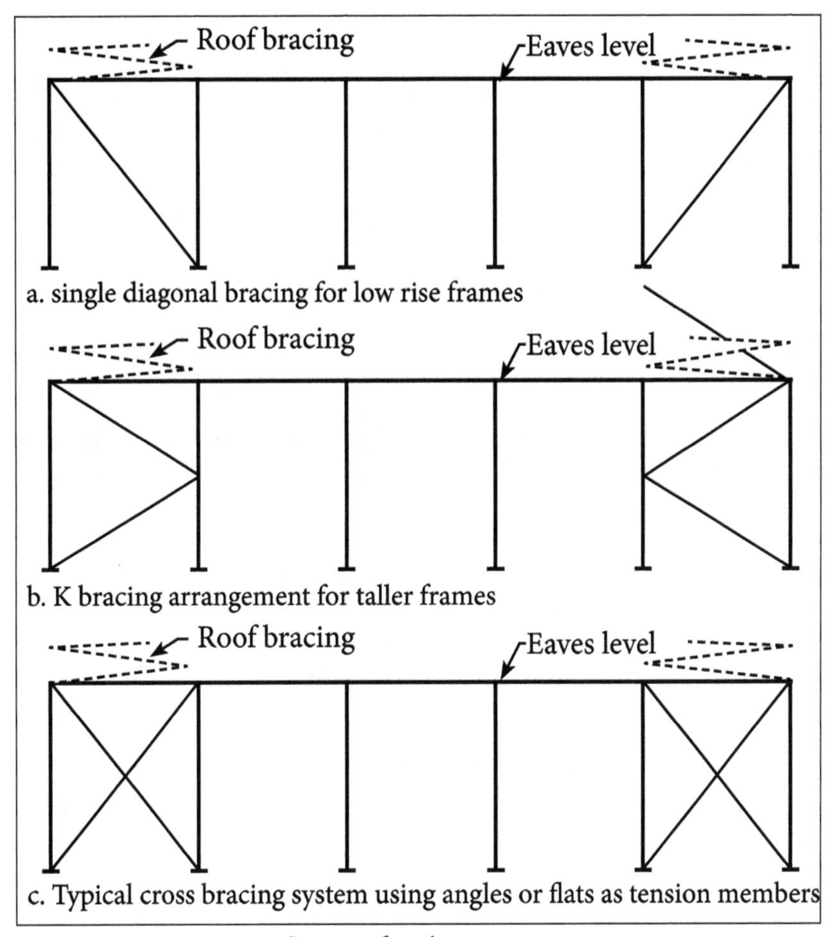

Common bracing systems.

8.1.2 Location of Bracing

The bracing may be located:

- At one or both ends of the building.
- Within the length of the building.
- In each portion between expansion joints (where these occur).

Where the side wall bracing is not in the same bay as the plan bracing in the roof, an eaves strut is essential to transmit the forces from the roof bracing into the wall bracing. An eaves strut is also required:

- To ensure the tops of the columns are adequately restrained in position.
- To assist in during the construction of the structure.
- To stabilise the tops of the columns if a fire boundary condition exists.

Portalised Bays

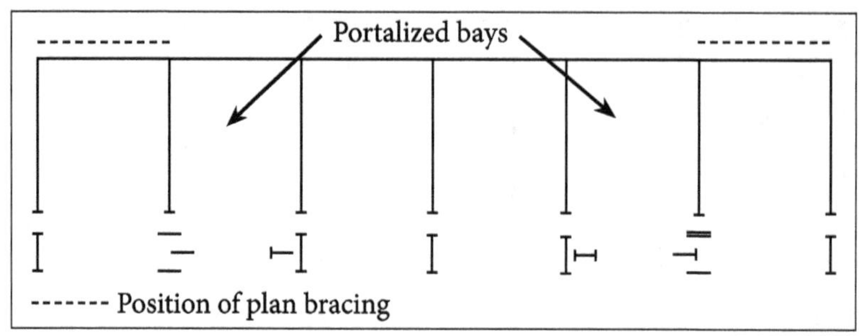

Longitudinal stability using portalised bays.

Where it is difficult or impossible to brace the frame vertically by conventional bracing, it is necessary to introduce moment-resisting frames in the elevations in one or more bays.

In addition to the general serviceability limit on deflection of h/300, where h is the height of the portalised bay it is suggested that:

- The bending resistance of the portalised bay (not the main portal frame) is checked using an elastic frame analysis.
- Deflection under the equivalent horizontal forces is restricted to h/1000, where the equivalent horizontal forces are calculated based on the whole of the roof area.

Bracing to Restrain Longitudinal Loads From Cranes

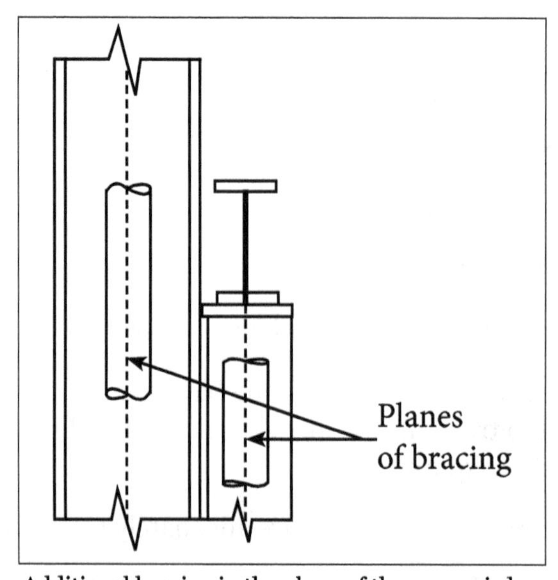

Additional bracing in the plane of the crane girder.

If a crane is directly supported by the frame, the longitudinal surge force will be eccentric to the column and will tend to cause the column to twist, unless additional restraint is provided.

A horizontal truss at the level of the crane girder top flange or, for lighter cranes, a horizontal member on the inside face of the column flange tied into the vertical bracing may be adequate to provide the necessary restraint.

For large horizontal forces, additional bracing should be provided in the plane of the crane girder.

Plan Bracing

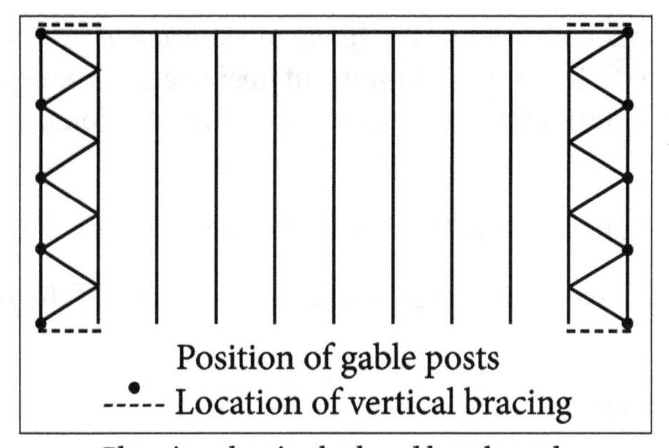

Plan view showing both end bays braced.

Plan bracing is located in the plane of the roof. The primary functions of the plan bracing are:

- To transmit wind forces from the gable posts to the vertical bracing in the walls.
- To transmit any frictional drag forces from wind on the roof to the vertical bracing.
- To provide stability during erection.
- To provide a stiff anchorage for the purlins which are used to restrain the rafters.

In order to transmit the wind forces efficiently, the plan bracing should connect to the top of the gable posts.

Restraint to Inner Flanges

Restraint to the inner flanges of rafters or columns is often most conveniently formed by diagonal struts from the purlins or sheeting rails to small plates welded to the inner flange and web.

Pressed steel flat ties are commonly used. Where restraint is only possible from one side, the restraint must be able to carry compression. In these locations angle sections of minimum size 40 × 40 mm must be used.

The stay and its connections should be designed to resist a force equal to 2.5% of the maximum force in the column or rafter compression flange between adjacent restraints.

Connections

The major connections in a portal frame are the eaves and apex connections, which are both moment-resisting. The eaves connection in particular must generally carry a very large bending moment.

Both the eaves and apex connections are likely to experience reversal in certain combinations of actions and this can be an important design case. For economy, connections should be arranged to minimise any requirement for additional reinforcement (commonly called stiffeners). This is generally achieved by:

- Making the haunch deeper (increasing the lever arms).
- Extending the eaves connection above the top flange of the rafter (an additional bolt row).
- Adding bolt rows.
- Selecting a stronger column section.

Typical Portal Frame Connections:

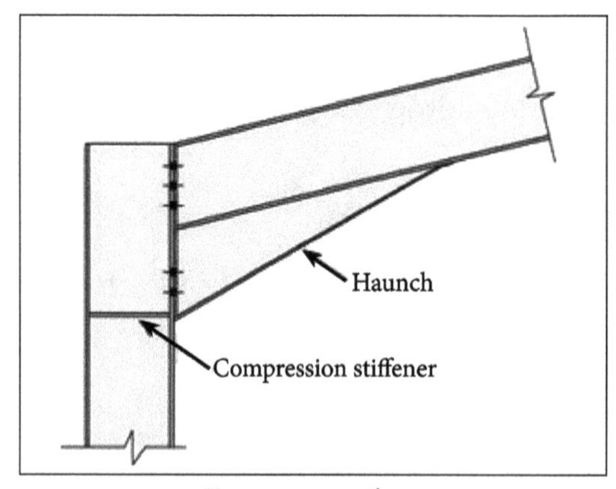

Eaves connection.

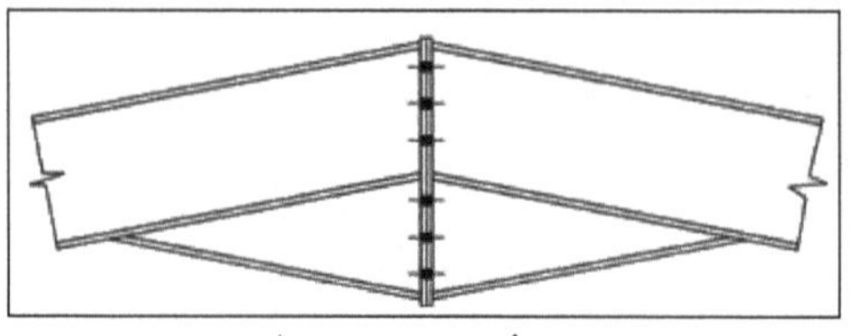

Apex connection.

Column bases:

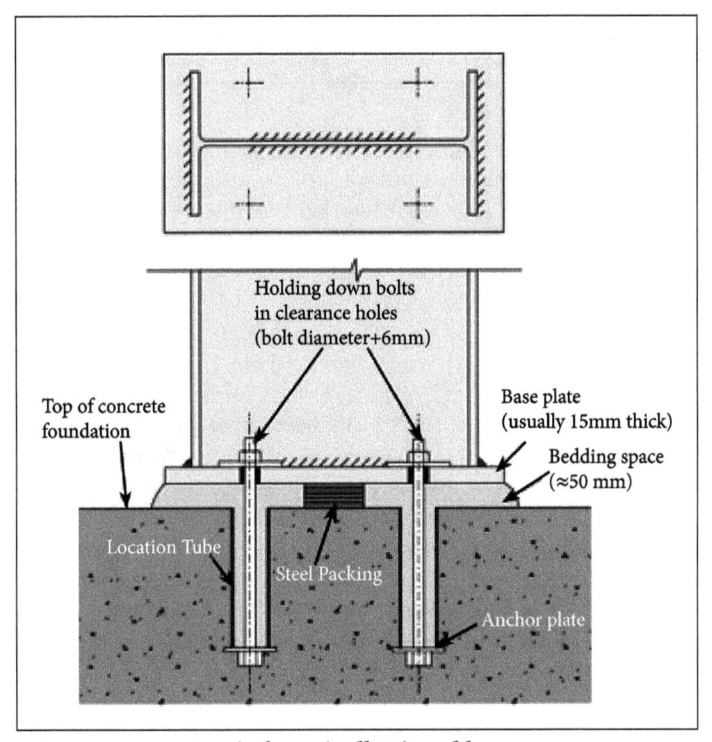

Typical nominally pinned base.

In the majority of cases, a nominally pinned base is provided, because of the difficulty and expense of providing a rigid base. A rigid base will involve a more expensive base detail, but more significantly, the foundation must also resist the moment, which increases costs significantly compared to a nominally pinned base.

If a column base is nominally pinned, it is recommended that the base be modeled as perfectly pinned when using elastic global analysis to calculate the moments and forces in the frame under ULS loading.

The stiffness of the base may be assumed to be equal to the following proportion of the column stiffness:

- 10% when assessing frame stability.
- 20% when calculating deflections under serviceability loads.

Permissions

All chapters in this book are published with permission under the Creative Commons Attribution Share Alike License or equivalent. Every chapter published in this book has been scrutinized by our experts. Their significance has been extensively debated. The topics covered herein carry significant information for a comprehensive understanding. They may even be implemented as practical applications or may be referred to as a beginning point for further studies.

We would like to thank the editorial team for lending their expertise to make the book truly unique. They have played a crucial role in the development of this book. Without their invaluable contributions this book wouldn't have been possible. They have made vital efforts to compile up to date information on the varied aspects of this subject to make this book a valuable addition to the collection of many professionals and students.

This book was conceptualized with the vision of imparting up-to-date and integrated information in this field. To ensure the same, a matchless editorial board was set up. Every individual on the board went through rigorous rounds of assessment to prove their worth. After which they invested a large part of their time researching and compiling the most relevant data for our readers.

The editorial board has been involved in producing this book since its inception. They have spent rigorous hours researching and exploring the diverse topics which have resulted in the successful publishing of this book. They have passed on their knowledge of decades through this book. To expedite this challenging task, the publisher supported the team at every step. A small team of assistant editors was also appointed to further simplify the editing procedure and attain best results for the readers.

Apart from the editorial board, the designing team has also invested a significant amount of their time in understanding the subject and creating the most relevant covers. They scrutinized every image to scout for the most suitable representation of the subject and create an appropriate cover for the book.

The publishing team has been an ardent support to the editorial, designing and production team. Their endless efforts to recruit the best for this project, has resulted in the accomplishment of this book. They are a veteran in the field of academics and their pool of knowledge is as vast as their experience in printing. Their expertise and guidance has proved useful at every step. Their uncompromising quality standards have made this book an exceptional effort. Their encouragement from time to time has been an inspiration for everyone.

The publisher and the editorial board hope that this book will prove to be a valuable piece of knowledge for students, practitioners and scholars across the globe.

Index